AF681488

ANALECTA HUSSERLIANA

THE YEARBOOK OF PHENOMENOLOGICAL RESEARCH

VOLUME I

A Humanities Press Book

We are pleased to send you a complimentary copy of

Analecta Husserliana. The Yearbook of Phenomenological Research. Vol.1

Anna-Teresa Tymieniecka, ed.

... *for review*

We would appreciate two copies of your review.

Publication date 1971

Price .. 15.50

HUMANITIES PRESS, Inc.

Publishers & Importers of Scholarly Books

303 PARK AVE. SOUTH NEW YORK 10, N. Y.

ANALECTA HUSSERLIANA

THE YEARBOOK OF PHENOMENOLOGICAL RESEARCH

VOLUME I

Edited by

ANNA-TERESA TYMIENIECKA

HUMANITIES PRESS / NEW YORK
D. REIDEL PUBLISHING COMPANY / DORDRECHT-HOLLAND

SOLE DISTRIBUTORS FOR U.S.A. AND CANADA
HUMANITIES PRESS / NEW YORK

Library of Congress Catalog Card Number 70–135105

SBN 90 277 0171 7

All Rights Reserved
Copyright © 1971 by D. Reidel Publishing Company, Dordrecht, Holland
No part of this book may be reproduced in any form, by print, photoprint, microfilm, or any other means, without written permission from the publisher

Printed in The Netherlands by D. Reidel, Dordrecht

FROM THE EDITOR

From 1913 till 1930 the *Jahrbuch für Philosophie und Phänomenologische Forschung*, edited by Edmund Husserl, served as the medium in which the most important advances in the new philosophical field of phenomenology saw the light of day. When in 1930 the *Jahrbuch* came to an untimely end, phenomenology had lost its central organ of communication.

As if to compensate for the ensuing lack of cohesiveness, phenomenology started to flourish in many new directions, and in countries that had not participated in its early development. In the three decades since the founder's death, phenomenology as a movement in active development has spread from Germany to Belgium, France, Poland, other European countries, the Americas and elswhere. But more important it has expanded in the variety of directions of thought and points of view. This dynamic expansion of phenomenology was stimulated by the activity of the Husserl Archives in Louvain, where under the directorship of Fr. Leo Van Breda Husserl's manuscripts were processed for publication and vividly discussed among scholars.

Thus phenomenology, undertaken by various thinkers, philosophers and scholars, each with his personal biases and tendencies, has penetrated the humanistic thinking of our epoch, but in turn has spread itself so thin that some of the most influential derivatives of phenomenology have only a remote connection with Husserl's own fundamental concerns. On the other hand, among these concerns that have had the widest influence we do not find some of his most essential insights.

In the most recent times, however, several new developments can be seen. The great wave of the phenomenological inspiration, having saturated the thirst for novelty and outlived its revolutionary stage, has passed away. After this float in retreat return to Husserl's original thought has become visible in several ways.

Firstly, the phenomenologico-existential trend chiefly represented by Heidegger, Merleau-Ponty and Sartre, which with its major theme of the life-world had during the two decades after the World War II dominated the phenomenological scene, is now approached critically and from a distance. Whereas the publication of a series of Husserl's own manuscripts has attracted attention to Husserl's own treatment of the constitution of the life-world. Hence a new tendency to disentangle the original phenomenological from the existential insights by returning to the source.*

Secondly, this reorientation of phenomenological interest stems also from the increasingly more critical attitude towards the hitherto dominant transcendental-subjective orientation with which phenomenology has been mostly identified in the last two decades. The quest after a change coincides with the translation of the monumental work of Roman Ingarden from his native Polish into German, which makes it at long last accessible to a larger audience. As it is well known, Ingarden with other Göttingen students of Husserl, did not follow him in his transcendental turn, in which Ingarden detected metaphysical assumptions that could compromise the claim of presuppositionlessness. He took up the first Husserl's project of phenomenology, which as a *scientia universalis* should for its start repose upon the eidetic reduction alone, and elaborated an *a priori* ontological foundation for this reduction.** Roman Ingarden's sudden death at the peak of his philosophical productivity, while this volume was already in proofs, means an irretrievable loss for the progress of phenomenological research.

Finally, having thus in front of us both of the primordial insights of Husserl (the eidetico-ontological and the transcendental) carried almost to their conclusions, we may turn to Husserl in a still more fundamental way. We may now try to pierce intuitively, philosophically *par excellence* the maze of phenomenological doctrines, approaches, theories, textual analyses and interpretations back to Husserl's own philosophical battle field. In this way we may question Husserl no longer as the proponent of a new approach but as a thinker fighting with *perennial puzzles presented to us by the universe and man.* On the basis of the scholarly apparatus of exegesis and interpretation of texts, so important a tool for the philosophical work, we may undertake *the reflection of phenomenology upon* itself, probing from within the phenomenological horizon of accomplish-

* Cf. the present writer's review of contemporary phenomenological literature: 'Dem Wendepunkt der Phänomenologie entgegen', in *Philosophische Rundschau* (1967).

** Cf. the present writer's 'The Second Phenomenology', in *For Roman Ingarden,* Nine essays in Phenomenology, M. Nijhoff, 1960.

ments into adequacy with which they come to grips with the evolving universe of cognition.

However as Roman Ingarden pointed out to me in a letter four weeks before his death, this matters chiefly for the sake of the progress of the original work itself. Indeed it is towards this aim that all the scholarly work and reflection have to be devoted.

In this 'back to the source' perspective a great need is felt for an international forum allowing communication of phenomenological research in progress among scholars. The demand for specialized scholarly contacts at regular intervals which would overcome the barriers of languages and geographical dispersion has found a positive response in the foundation in April 1969 during the International Phenomenological Conference at Waterloo, Ontario of *The International Husserl and Phenomenological Research Society*, which encouraged by the Waterloo exchange will meet every second year in different countries of both continents.

However, the phenomenological research remains scattered in different periodicals, in different languages, difficult to trace. Each periodical imposing its policy, conditions and limits upon the style of presentation, length of paper etc. hampers the essential research value and its communication. Phenomenological research in its actual turn calls, on the contrary, for and seems to contain the promise of a fundamental nucleus of research and a community of minds to emerge.

The reviving of Husserl's own *Jahrbuch* in the present *Analecta Husserliana, The Yearbook for Phenomenological Research*, which will appear periodically with essays in three languages (English, French and German), allowing for an essay length up to 100 pages is devised to fill this need and serve this purpose. Avoiding any doctrinary claim or policy it is open to all competent, specialized phenomenological inquiry, which can be basically traced by its relation to its source in Husserl. May it contribute towards establishing a second phase of fundamental continuity of work in the field of phenomenology!

ANNA-TERESA TYMIENIECKA

TABLE OF CONTENTS

ANNA-TERESA TYMIENIECKA

DIE PHÄNOMENOLOGISCHE SELBSTBESINNUNG

I

*Der Leib und die Transzendentalität in der gegenwärtigen phänomenologischen und psychiatrischen Forschung**

Von den beiden Richtungen der Phänomenologie, der eidetisch-strukturellen und der transzendentalen, scheint diese letzte bis jetzt nicht nur eine vielseitige sondern auch eine fast endgültige Entwicklung gefunden zu haben. Es scheint als ob jetzt, nachdem diese Richtung der Phänomenologie sich vielseitig ausgewirkt hat und wir die Ergebnisse vergleichen können, der Zeit reif wäre um durch die grossartig entwickelten Analysen, Themen, Doktrinen hin zu den philosophischen Problemen durchzudringen mit welchen sie grundsätzlich beschäftigt sind und zu sehen was eigentlich zu ihrer Lösung beigetragen hat, und was verfehlt wurde, aber doch im phänomenologischen Rahmen erziehlt werden kann. In dieser Perspektive wollen wir uns nachstehend mit dem Problem der Transzendentalität und des Leibes befassen.

Wie bekannt, die fundamentalen Ansätze zu der grundlegenden Bewusstseinsforschung, die die neuzeitliche transzendentale Phänomenologie in ihrem Anspruch auf unmittelbare Evidenz und Autonomie ihres Untersuchungsgebietes überhaupt bestimmt, stammen von der Philosophie Franz Brentanos. Die wesentlichen Richtlinien dieser Philosophie wurden aber von Husserl und seinen Nachfolgern vollkommen umgewandelt. Diese Umwandlung betrifft insbesondere die Rolle des Bewusstseins inmitten des totalen Naturgefüges. Von der bei Brentano zentralen und autonomen Stellung des Psychischen inmitten des gesamten Weltbildes, ist Husserl zu dem Extrem eines vollkommen in sich selbst ruhenden, selbständigen, *absoluten* – in sich selbst 'wurzelnden' reinen Bewusstseins fortgeschritten.

Dieser Bewusstseinsbegriff kulminiert in einer, auf den konstitutiven Leistungen des transzendentalen Bewusstzeins allein ruhenden, Weltauffassung. Die Körperlichkeit der Natur wird als durch die Konstitution gestaltet mit in diese Weltauffassung einbegriffen. Das transzendentale

Tymieniecka (ed.), Analecta Husserliana, Vol. I, 1–10. *All Rights Reserved*
Copyright © 1970 *by D. Reidel Publishing Company, Dordrecht-Holland*

Bewusstsein selbst gilt als 'verleibt', d.h. die konstitutiven Leistungen erstrecken sich bis zur sinnlichen Wahrnehmung als *Urmodus* der Anschauung in welcher der Leib als Erfahrungsinstrument erscheint.[1] So aufgefasst, ist der Leib nicht ein Glied des Naturgefüges in dem Gesamtbild des Wissens, sondern dessen Urheber in dem System der transzendentalen Konstitution, in welchem System er als ein Grenzelement fungiert, als der Nullpunkt des gesamten Weltbildes.

Die Verabsolutierung des transzendentalen Bewusstseins scheint, wie es auch scharf in der neuesten Husserlforschung und Diskussion zum Vorschein kommt, ernste, vielleicht sogar unüberbrückbare Schwierigkeiten mit zu bringen. Es scheint als ob diese Schwierigkeiten zuletzt in der Auffassung des Leibes zu dem Nullpunkt der Konstitution reduziert gründen werden. Es bestehen zweifellos Schwierigkeiten des Problems der Beziehung des transzendentalen und natürlichen Bewusstseins, jene der Rolle des *a priori* in der Konstitution, sowie das noch zu erforschende Problem der *ursprünglichen Spontaneität* des transzendentalen sowie des natürlichen Bewusstseins[2], die alle auf die Auffassung des Leibes zurückzuführen sind. (Merleau-Ponty, der diese Schwierigkeiten offenbar gesehen hat und zu vermeiden versuchte scheint sie nicht gelöst zu haben indem er, letzten Endes, zu einem primitiven Seinsbegriff gelangt.)

Hingegen in dem Gedankensystem Brentanos, in welchem die Husserlsche Phänomenologie ihre Inspiration fand, spielt die Auffassung des Leibes eine doppelte Rolle, dem Bewusstsein und der körperlichen Natur gegenüber.

Die zentrale Stellung des Bewusstseins im Rahmen der stufenweise aufgefassten Gesamtheit des Wissens ruht, auf der einen Seite, in der einzigartigen Natur der Bewusstseinsforschung die die selbstevidenten psychischen Vorgänge und ihre autonomen gegenständlichen Inhalte in einer *deskriptiven* Analyse behandelnd, zu den letzten objektkonstitutiven Elementen vordringen soll und dadurch eine allumfassende Auslegung des Wissens (eine *scientia universalis*) beabsichtigen konnte. Als solche fungiert sie als Fundament der Philosophie.

Auf der anderen Seite aber, erstreckt sich das Bewusste, durch die besondere Art seiner 'Verleibung', stufenweise – und seine spezifische Autonomie bewahrend – in das totale, auch als autonom aufgefasste, Naturgefüge. Das Insspieltreten der psychischen Vorgänge überhaupt, ihre Sukzession, Verflechtung, Motivierung u.dgl., d.h. das 'Leiblichbewusste' ist in dem 'Leiblich-naturlichen' verwurzelt. Es scheint nun dass in diesem Gedankenzusammenhang die letzten Gesetze des Bewusstseinshorizontes als der universalen Wissensauslegung, in der vermittelnden Rolle des Leibes begründet werden. Das heisst, um die letzten Gesetze des Gegenständlichen-oder-Erlebnisbewussten zu gewinnen würde man eine genetisch-physiologische Forschung zu berücksichtigen haben. Die Ergebnisse der wissenschaftlichen Naturforschung würden dann, entsprechend bewertet, in ihrem Fortschritt als eine wichtige Erkenntnisquelle der Philosophie zu Rate gezogen.

Angesichts des grossartigen Fortschritts der Naturwissenschaften der in der Gegenwart erzielt wird, gibt uns diese ursprünglich der Phänomenologie vorausgehende Einstellung zu denken. Die Frage, die nun entsteht und die hier aufgeworfen sowie behandelt wird, ist: Wie sich das Problem Leib-Bewusstsein im Lichte der neuesten Naturforschung, insbesondere der psychiatrischen Forschung zeigen würde. Psychiatrie vereinigt auf eine einzigartige Weise die konkrete Natur- mit der Bewusstseinsforschung und hat an dem riesigen Fortschritt der Naturwissenschaften seit Brentanos Zeit teilgenommen. Zumal wurden einige Richtungen der psychiatrischen Forschung durch die Annahme fundamentaler phänomenologischer Einsichten auf neue Wege umgestellt.[3] Die ursprüngliche phänomenologisch inspirierte Richtung der Psychiatrie hat nicht nur die Selbstevidenz und die Autonomie der Bewusstseinsvorgänge und ihrer Inhalte, sondern auch die transzendentale Konstitution der Lebenswelt (in der Husserlschen-Heideggerschen Auffassung) angenommen und dabei auch die Auffassung des Leibes als einen Nullpunkt der Konstitution mit in Kauf genommen. Nun aber scheint (in ein paar Jahrzehnten dauernden Forschung) die jüngere Generation der phänomenologisch inspirierten Forscher zu den Grenzen ihrer Gültigkeit selbst durchzudringen.

Die phänomenologische Bewusstseinsforschung, die bis jetzt in dem Enthusiasmus der Entdeckung des Feldes der Bewusstseinsforschung sich hauptsächlich der *Regelmässigkeiten* der psychischen Vorgänge gewidmet hat und von diesen Regelmässigkeiten auf die Natur der Konstitution reflektierte, kann vielleicht tiefere Einsichten in die endgültige Lage der Konstitutionsproblematik von dem Studium der *Unregelmässigkeiten* gewinnen, die pathologische Fälle (durch die Psychiatrie behandelt) und zeigen.

Im Lichte der neuesten psychiatrischen Einsichten in Gegenüber-stellung zum Erbgut der Phänomenologie, drängen sich zwei Gruppen von Fragen auf, die Gegenstand unserer Untersuchung sind.

Die erste Gruppe der Fragen würde die *Absolutheit der konstitutiven Vermöglichungen* betreffen. Indem wir nämlich die pathologischen Unregelmässigkeiten der Bewusstseinsvorgänge auf eine phänomenologisch reduzierte Weise betrachten, entsteht die Frage, ob das transzendental konstitutive Bewusstsein vielleicht nur eine fragmentarische – wenn auch ausschlaggebende – Rolle in der Pluridimensionalität des bewussten Seins spielen würde. Ihre 'absolute' existenziale Stellung als in sich selbst wurzelnd und aus eigener Spontanität die bewussten Vorgänge selbst begründend wird fragwürdig.

Dadurch kommen wir zu der zweiten Gruppe der Fragen, die die *Natur und Rolle des Leibes in der Konstitution* betreffen. Vielleicht ist es nicht genügend zwischen dem Leib als miterfahrenden Instrument der Wahrnehmung und dem Körper als dessen Vergegenständlichung allein zu unterscheiden. Vielleicht weist die Pluridimensionalität der Bewusstseinsvorgänge auf eine dritte, diesmal ausschlaggebende Auffassung des Leibes, die dem über die transzendentale Konstitution erweiterten Rahmen des Bewussten genugtun könnte.

I. DAS SELBSTGEGEBENE

Indem man zwei Ebenen des transzendental-konstitutiven Bewusstseins (im Husserlschen Sinne) als die des *aktuellen Feldes*, wo Erlebnis überhaupt zustandekommt, und die der *Selbstkonstitution* des Ich als Person in ihrer Lebenswelt unterscheidet, würde sich die Absolutheit des konstitutiven Bewusstseins bewähren, wenn nicht nur die konstituierten Phänomene den Regeln des konstitutiven Systems entsprechend gestaltet würden – was unbestritten bleiben mag –, aber auch wenn *alle* im aktuellen Felde des Bewusstseins und in der Ichgestaltung (als *Person* in ihrer Welt) auftretenden Erlebnisformen und Bedeutungen *gesetzmässig* und *ursprünglich* in Bezug auf die konstitutive Genesis mit ihrem Horizont expliziert sein könnten. Aber die durch das Miteinbeziehen des 'pathologischen' Erlebnisses erweiterte phänomenologische Forschung bringt dem 'regelmässigen' (die Konstitution mit ihrem Gestaltungssystem als Norm angenommen), das pathologische Erlebnis gegenüber und weist auf beiden Ebenen (d.h. der Erlebnis-sowie-Ichgestaltung) Bildungen auf die ausserhalb der Gesetzmässigkeit und Fähigkeiten zum Ins-sein-bringen des transzendentalen Bewusstzeins zu liegen scheinen.

(a) In der genetischen Selbstbildung des Ich als Person (das Ins-sein-treten der Individualität/Freiheit der Anonymität der Natur gegenüber) sind die Regelstrukturen der Umbildung und Integration der Erlebnisse (*expérience vécue*) entscheidend.[4] Nun aber weist die phänomenologische Analyse der äusseren Wahrnehmung in 'pathologischen' Fällen (z.B. der Phobien), verglichen mit der 'normalen' (d.h. transzendental-konstitutiven) Wahrnehmung desselben Gegenstandes, Unterschiede auf, die das Noema selbst betreffen.[5] Während jede Änderung, die in der 'normalen' Wahrnehmung auftritt, sich prinzipiell in den Rahmen des transzendental-konstitutiven Systems genetisch rückverfolgen und explizieren lassen soll, lässt sich die spezifisch 'pathologische' Abänderung des Noema nicht auf die Bahn der regelmässig konstituirenden Wahrnehmung zurückzubringen: sie ist ihr fremd und scheint dagegen zu der spezifisch 'umgewandelten' Lebensweise der Person, deren Erlebnis es ist, zu gehören. Diese umgewandelte Lebensweise der Person aber ist selbst nur durch das Besondere der einzelnen Erlebnisse im aktuellen Feld des Bewusstseins leibhaftig gegeben; darüber hinaus lässt sich weder ihr, dem konstitutiven Ichprozess gegenüber, sich Einschliessen (Integration), noch ihr einzigartiger Ursprung anhand des konstitutiven Horizontes explizieren.

Analysen dieser Art scheinen folgende Hinweise zu enthalten: (1) es scheint als ob die im aktuellen Felde des Bewusstseins leibhaftig auftretenden Erlebnisinhalte nicht ausschliesslich der stiftenden Gesetzmässigkeit des transzendental-konstitutiven Bewusstseins entspringen, sondern dass das Bewusstseinsfeld auch Erlebnisse aufweist, die von ausserhalb des konstitutiven Systems ihren Ursprung (Motivation) sowie Gestaltung beziehen; (2) diese Unzulänglichkeit der transzendental-konstitutiven Vermöglichkeiten auf den Sachbestand deutet, dass diese *nicht über den ganzen* Umfang des bewussten Lebens verfügen würde; (3) zuletzt, das Auftreten im aktuellen Bewusstseinsfelde von konstitutiv-fremden Gestaltungen der Erlebnisse auf eine ausserkonstitutive Bewusstseinsebene weist, deren Ursprung- und Gestaltungsfähigkeiten in besonderen Fällen als Urheber und Fahrzeug der Erlebnisse fungieren.

(b) Diese Fragen präzisieren sich weiter in der phänomenologischen Untersuchung des '*aufgelösten Bewusstseinsfeldes*' (*déstructuration du champ de la conscience*).[6] Das Feld des aktuellen Bewusstseins, das durch die Bewirkung der transzendentalen Konstitution selbst entsteht als die grundlegende Ebene des Emporsteigens der Individualität/Freiheit wo die Aktualität des individuellen Erlebens überhaupt zustandekommt, erscheint in den Erlebnissarten mancher Geisteskrankheiten vollkommen aufgelöst. Die Art und Weise dieser Auflösung lässt sich phänomenologisch anhand der leibhaft auftretenden Erlebnisse ergreifen und verfolgen und bringt die spezifische Dreidimensionalität des Feldes selbst sowie die eigenartige Rolle dieser konstitutiven Leistungsfähigkeiten zum Vorschein. (1) Das Einordnen der Erlebnisse des konstitutiv organisierten Feldes des aktuellen Bewusstseins in das System der geschichtlichen Existenz wird aufgehoben in Fällen wo das emotionelle Gleichgewicht (zwischen der statischen Tendenz der Angst und der dynamischen Tendenz des Begehrens) nicht erhalten wird. Die zeitliche Phase der Gegenwart konstituiert sich nämlich nicht in ihrer transzendentalen Struktur, indem die Erlebnisse aus der konstitutiven Ordnungslinie 'herausfallen'. Anstatt sich dem realitätsetzenden System einzuordnen, erscheinen sie als 'Irrealitäten'. Indem in der Auflösung des Feldes die konstitutive Zeitordnung als Realitätsordnung sich erweist, weisen die Erlebnisse, die als Irrealitäten auftreten auf eine ausserkonstitutive, eigenmässige Motivation (Ursprung und Gestaltungsmotivation) hin. (2) Die Ordnung der räumlichen Gestaltung, die die transzendental-konstitutive Polarität Subjekt-Objekt (die spezifisch konstitutive Intentionalität) selbst mit sich führt wird in Fällen der 'morbiden Halluzinationen' aufgelöst. Die Fragment-Gestalten die das Erlebnis zur Schau bringt, stehen dem Subjekt als Pole der Akte sowie als geschichtliche Individualität fremd gegenüber und weisen dadurch auf eine aussersubjektive, unindividuelle Quelle hin. (3) Zuletzt, sogar die konstitutiven Strukturen des Sich-überhaupt-der-Welt-Öffnen, der Realitätssetzung selbst, die das Erlebnis als 'wirklich' stiftet, scheinen sich in den Zuständen der 'Traumverwirrung' (*états confuso-oniriques*) nicht zu verwirklichen. Im Gegenteil, die Objekt-polarisierende Intentionalität kommt überhaupt nicht zustande, da die Erlebnisse als rein fiktive, symbolische Traumerscheinungen von der Ichheit nicht abgesondert sind. Ihre Intentionalität weist auf ein unindividualisiertes Phantasiereich als ihre Herkunft hinaus.

Die Ergebnisse dieser Auseinandersetzung der Gestaltungsformen, die im Bewusstseinsfeld auftreten, zusammenfassend lässt sich behaupten, dass:

(1) nebst der dreidimensionellen Erlebnissphäre des Bewusstseinsfeldes als diejenige der Individualität/Freiheit, sollen wir als ihre Vorstufe die Bewusstseinsebene der kollektiven Phantasiegebilde annehmen. Obwohl diese Ebene im Bewusstseinsfeld nur als ein *Bezugsystem* 'intendiert' wird, erscheint dieses aber als ein *notwendiges Korrelat des individualisierenden Funktionen des Bewusstseins.*

(2) Das Auftreten im Bewusstseinsfelde anderer als die konstitutivbedingten Erlebnissformen zeigt, dass die *bildende Spontanität* des bewussten Lebens keineswegs auf die des konstitutiven Systems begrenzt sein kann. Im Gegenteil, das Auftreten des individualisierenden Erlebnissystemes überhaupt – der Anonymität der 'phantastischen' Intentionalität gegenüber – weist darüber hinaus auf eine *Uranlage des Bewusstseins* die über das Sortieren und Wahl der Gestaltungssysteme, und das Insspiel treten der konstitutiven Fähigkeiten überhaupt entscheiden würde.

II. DIE BEZUGSYSTEME DES BEWUSSTEN

Unsere Ergebnisse die die absolute Gültigkeit des transzendental-konstitutiven Systemes zu begrenzen scheinen bringen Fragen auf, die im Rahmen der phänomenologischen Bewusstseinsuntersuchung nicht mehr beantwortet werden können. Diese Untersuchung enthält aber Ansatzpunkte und Richtlinien zu deren Behandlung. Es zeigt sich nämlich das die spezifisch phänomenologische Auffassung des Bewussten als ein, seit Brentano über Husserl, die den *dynamisch-bezughaften* Aspekt des Bewussten als seine hauptsätzliche Natur emporzuheben scheint, wurde in den obigen Untersuchungen nicht nur betont, sondern sogar weitererstreckt, indem das 'leibhaftig als Phänomen gegebene', in seinem erfassbaren noetisch-noematischen Gehalt, sich systematisch ordnende Hinweise enthält auf die ihm als Korrelat zuzuordnenden und von ihm aus als *Bezugsysteme* zu erfassenden weitere Stufen im Gesamtentwurf des menschlichen Seins, die sich in der transzendentalen Konstitution auch bewirken würden.

Wie betont, die psychiatrische Forschung greift diese Fragen auch von der Seite des Nicht-bewussten, der Natur an. Die psychiatrische Diagnostik und klinische Therapeutik, in welcher die phänomenologische Deskription der Erlebnisweisen mit den Funktionsstrukturen des Organismus konfrontiert wird, weist mit steigender Eindeutigkeit Korrelationen zwischen den Strukturen der Funktionssysteme des Organismus die sich auf verschiedenen Stufen dessen Gestaltung bewirken und jenen des bewussten Lebens, auf. Die anatomische Struktur des Gehirns, die Systeme des Gehirnbewirkens werden in eindeutige Korrelationen mit der Gestaltungsweise und Motivation des Erlebnisbewusstseins gebracht. Darüber hinaus, die Erfolge der klinischen Anwendung der pharmazeutischen Mittel in Geisteskrankheiten scheinen dieses weiter in Korrelationen zwischen der Dynamik und Gestaltung des Bewussten, auf der einen Seite, und der Bewirkung der anorganischen Natur in physiologischen Prozessen des Lebendigen, auf der anderen Seite, zu erweitern.[7]

Die Frage nun entsteht, ob wir die von der Seite der Naturforschung aus aufgewiesenen Funktionssysteme des Organismus die als sich in dem bewussten Leben bewirkende klinisch festgestellt werden, nicht mit den durch die phänomenologische Bewusstseinsuntersuchung aufgewiesenen Bezugsystemen des Bewussten zusammenbringen sollen. Sollen wir nicht die durch das phänomenologisch erfasste Erlebnis ihr als Korrelat zugehörigen Bezugsysteme als den *reduzierten* Ausdruck der naturwissenschaftlich erfassten und objektivierten Funktionssysteme des Organismus betrachten? Was in der naiven 'vorkritischen' Einstellung der Naturwissenschaft als kausales sich-Bewirken erscheint, zeigt sich von der Phänomenologie des Bewussten aus differenziert in mehrzähligem Korrelationsgefüge der *gestaltenden Spontanität*, die auf verschiedenen Gestaltungsstufen eine andere Form annimmt; die dynamisch gestaltenden Stufen selbst bleiben *autonom* in ihrem Bereich des sich-Bewirken stehen aber in einem stufenförmigen *Motivationsgefüge* das über die Ursprünglichkeit und die Gestaltung entscheiden.

Die *Uranlage* des Bewussten, auf welche das in seinem Bereich autonome, transzendental-konstitutive Bewusstsein grundsätzlich angewiesen zu sein scheint, würde sich nun auf dieses stufenweise organisierte dynamische Bezugsgefüge erstrecken. Sie würde in ihrer spezifischen Gestaltung inmitten dessen gesucht werden und sich genau so über das Nicht- und Vorbewusste (Unbewusste) wie über die verschiedenen Dimensionen der bewussten Entfaltung erstrecken, ohne dass man im Rahmen des Gesamtentwurfes des menschlichen Seins zwischen dem Bewussten und Nichtbewussten eine scharfe Scheidungslinie zu ziehen vermag.

III. DER LEIB ALS DIE URANLAGE DES BEWUSSTSEINS UND ALS EIN KONVERGENZZENTER IM GESAMTGEFÜGE DER NATUR

Auf diesem Umwege sind wir zu dem Problem des Leibes gekommen. Bis jetzt scheinen die obigen Ausführungen auf einen Rückkehr zu dem stufenförmigen Gesamtbild des Wissens wovon die Phänomenologie entsprungen ist, zu verweisen. Dieses Gesamtbild nun ist zweifellos bedeutend umgedeutet erschienen. Von ihm braucht man nur einen Schritt weiter zu gehen um unsere Erörterungen für die Stellung des Leibproblems zu bewerten.

Wie Husserl es ausdrücklich betont, und seine Nachfolger es aufgenommen haben, die transzendentale Genesis als ein bedeutungsstiftendes Emporsteigen des sich der Welt öffnenden Ich hebt ursprünglich mit der sinnlichen Erfahrung ab. Der Leib wird ursprünglich als ein System der sinnlichen Wahrnehmungsorgane jeder sinnlichen Erfahrung entsprechend mitgegeben. In der Perspektive unserer skizzenhaften Analysen der *Konvergenzzenter des dynamischen Bezugsgefüges*, der, auf der einen Seite, als die Uranlage des Bewusstseins zu fungieren scheint, auf der anderen Seite ein Organisationsnetz der organischen Bewirkungen inmitten des ausserindividuellen Natursystems darstellt, scheint diesem Leibsystem, das in der konstitutiven Analyse der Wahrnehmung sich kundgibt, auf einer anderen Ebene zu entsprechen.

Indem wir jetzt an die Auffassung des Leibes als den Nullpunkt der Konstitution anknüpfen, bekommt diese, in dieser neu sich öffnenden Perspektive eine besonders starke, aber auch bedeutend verschiedene Prägung. In seiner Bewusstseinszuwendung wäre der Leib der Urheber des sich stufenweise entfaltenden bewussten Lebens, das in der individualisierenden und Freiheit stiftenden Rolle der Stufe des konstitutiven Bewusstseins kulminieren würde. In der entgegengesetzten Richtung, die der starren Natur, scheint der Leib der Konvergenzzenter der Naturprozesse, die einem besonderen Plan, den er entwirft, folgend, in ihm zusammenfliessen, zu bedeuten.[8] Ausführliche Untersuchungen sind natürlich nötig um die *dynamischen* sowie *ordnenden Funktionsverhältnisse* dieses Gesamtsystems, das in dieser merkwürdigen Spannung Natur-Geist kulminiert, aufzuklären. Unser gegenwärtiges Vorhaben war es, den Weg, den die Entwicklung der Phänomenologie selbst zu betreten scheint, zu bahnen. Auf diesem Wege, hat die Autonomie des konstitutiven Bewusstseins – die Hauptthese der Phänomenologie – sich bewährt, auf der anderen Seite aber, wir sind von der Phänomenologie her zu dem *selbsteigenen Seinsrecht* der Natur durchgedrungen, und vor allem, aber, sind wir vielleicht erst jetzt imstande die letzten Bedingungen, die *Uranlage* des Bewusstseins selbst zu erforschen.[9]

Man soll es aber nicht verhehlen, dass in dieser neuen Perspektive in welche die Phänomenologie, sich selbst nachprüfend and bewertend, eintritt, und die hier nur skizzenhaft entworfen ist, umzählige Fragen und Schwierigkeiten sich aufdrängen. Die Fragen nach den Urprinzipien, die die Organisation des Bezugsgefüges des Bewusstzeins leiten, nach dem vielseitigen System des Leibes, der das Bewusste mit dem Nichtbewussten in einer einheitlichen Entfaltung zusammenbringt, letztlich die Frage nach der Wahl der Gestaltungsformen der Konstitution sowie des menschlichen Wesens selbst, öffnen die Pforte der Metaphysik.

ANMERKUNGEN

* Eine Variation dieser Abhandlung wurde im Rahmen des Symposion 'Brentano und die Phänomenologische Psychologie' des XIV. Intern. Phil. Kongresses in Wien 1968 vorgetragen.

[1] V.a. cf. auch *Die Krisis der Europäischen Wissenschaften und die transzendentale Phänomenologie*, Martinus Nijhoff, III, 28, 1954, S. 107–**11**.

[2] Cf. den Standpunkt den ich vertrete in 'Dem Wendepunkt der Phänomenologie entgegen', *Philosophische Rundschau* **14** (1967), Heft 2/3.

[3] Cf. Anna-Teresa Tymieniecka, *Phenomenology and Science in Contemporary European Thought*, The Noonday Press, New York, 1961.

[4] Cf. Henri Ey, *La Conscience*, *P.U.F.*, Paris, 1963.

[5] G. Lanteri-Laura, 'Les problèmes de l'inconscient et la pensée phénoménologique', in *L'inconscient*, Desclée de Brouwer, 1966.

[6] Henri Ey, *La conscience*.

[7] H. Azima, *Problèmes biophysiques de la conscience*, 1954; Cl. Blanc, *Conscience et inconscient dans la pensée neurobiologique actuelle*, 1966.

[8] Cf. meine kosmologische Konzeption des realen individuellen Seins in *Why is there Something rather than Nothing. Prolegomena to the Phenomenology of Cosmic Creation*, Van Gorcum, Assen, Niederlande.

[9] Cf. Anna-Teresa Tymieniecka, 'Cosmos, Nature and Man and the Foundations of Psychiatry' im *Heidegger and the Path of Thinking*, Duquesne University Press, 1970.

JOSEPH J. KOCKELMANS

WORLD-CONSTITUTION

Reflections on Husserl's Transcendental Idealism

I. INTRODUCTION

For the series of lectures given in 1907 in which he made his first attempt to present a brief outline of his phenomenology,[1] Husserl took his starting point in a comparison of the 'natural' and the 'philosophical' sciences. For him there is a radical difference between these two types of sciences, and this difference is ultimately rooted in an essential disparity between the 'natural' and the 'philosophical' attitudes. Typical of the 'natural' attidue, which is characteristic of our pre-scientific life as well as of the sciences, is that the possibility of our knowledge is taken as a self-evident fact. Mainly because of the successes attained by the sciences there is no need felt to raise the question concerning the possibility of our knowledge. Evidently this is not to say that in his pre-scientific life man does not reflect upon his knowledge. For, obviously, it is possible to make inquiries about human knowledge and to show, for instance, that knowledge essentially is knowledge of an object and that it is this because of a meaning which is inherent to it. But in so doing man always remains within the realm delineated by the general thesis of the natural attitude in which the very possibility of knowledge is and remains taken as a self-evident fact.[2]

In adopting a completely different attitude, Husserl says, one can conceive of knowledge as a psychological experience of a certain subject. Then the question may be asked: How can this subject be *certain* that his knowledge agrees with the things it refers to? At this point the self-evident givenness of the objects of our natural attitude and of the world itself becomes problematic to him. How do I, this particular subject, know that my subjective experiences are not all there is? How do I know that there are real things 'outside' which are independent of my experiences? How can I be certain that through my knowledge of them I know things as they are 'in themselves'?

The point where all problems concerning the possibility of knowledge

Tymieniecka (ed.), Analecta Husserliana, Vol. I, 11–35. *All Rights Reserved*
Copyright © 1970 *by D. Reidel Publishing Company, Dordrecht-Holland*

and the age-old problems around it converge, is found, in Husserl's opinion, in epistemology as well as in metaphysics. Epistemology has a double task: (1) it must show that the errors into which a natural reflection upon the relation between knowledge and the objects known must almost necessarily fall, can in reality be avoided and that, therefore, no sceptical theory about the essence of man's knowledge may be accepted; (2) by accurately examining the very essence of knowledge epistemology must try to find a solution for the problems concerning the real relations between knowledge, its meaning, and its object.[3]

By attempting to answer these questions, epistemology becomes a critique of our natural knowledge and of all the sciences of the natural attitude, enabling us to arrive at an interpretation of the value of the results of the sciences. Its aim is to put an end to all arbitrary interpretations of the natural sciences. Thus epistemology, in which the difference between philosophy and science reveals itself, must show that the natural sciences cannot pronounce definitive judgments about Being and that such judgments can be formulated only by a metaphysics which emanates from a 'critique' of the sciences.[4]

But if it is true that epistemology must explain the essence of knowledge and of the objects known, then one may say that epistemology and phenomenology of knowledge are identical, and also that epistemology as just described, is the fundamental part of the whole of phenomenology. In other words, phenomenology is a complex of philosophical disciplines, of which the phenomenology of knowledge (epistemology) constitutes the foundation.[5]

Elaborating this conception of phenomenological philosophy Husserl begins by pointing out that phenomenology will never reach a radical foundation if it does not limit itself to a study of the 'adequately self-evident', the 'apodictic evident', or what amounts to the same, the 'genuinely immanent'. In other words phenomenology may not use that which is transcendent, which is not genuinely immanent.[6] Thus it must take its starting point in a phenomenological reduction and exclude all that is transcendently posited. For if I do not know how knowledge can reach that which is transcendent, that which is not given 'in itself' but is merely 'intended as being out there', no knowledge or science of the transcendent can help me to reach a radical point. It is apparent that in episytemolog nothing is gained by reducing things to my experiences of them. For a concrete man and the concrete things he knows are not absolute data and, thus, his 'mental activities' are no absolute data, either. The phenomenological reduction, in other words, has nothing in common with Hume's reduction of things to sense impressions. Such a conception of the reduction would necessarily lead to psychologism. The question is not: How can I, this particular man, reach in my experiences as mental processes something existing in itself, out there and beyond me? But the question here is rather: How can the absolute givenness of knowledge reach something not self-given, and how must this 'reaching beyond itself' precisely be understood?[7]

After we have secured for ourselves by means of the reduction the realm of phenomena, we must take a second step which, within the domain of the apodictically evident, leads us from the concrete and contingent to the domain of the universal and necessary, that is to say to the realm of the *a priori.* This second step is what Husserl calls 'eidetic abstraction' or 'eidetic reduction'.[8]

These two steps, however, do not suffice to clearly delineate what phenomenology of knowledge really seeks to accomplish. For if the subject matter of phenomenology is found in the domain of the universal and necessary, then it is clear at once that the identification of the absolutely given with the genuinely immanent is going to create problems. For that which is universal may be absolutely given, but it can never be genuinely immanent. The act of knowledge in and through which I know the universal is something singular, whereas the universal itself which is given in evidence is nothing singular but is precisely a universal which as such is transcendent in the genuine sense. To solve this problem one must further determine the meaning of the phenomenological reduction. The reduction does not exclude all that is genuinely transcendent, but merely the transcendent as something to be accepted as existing. In other words, the domain of phenomena in which phenomenology is interested is the field of absolute cognitions within which the ego, the world, God, and whatever else may be a legitimate subject matter for science must be held in abeyance, in other words cognitions which are not dependent upon these matters taken as realities in the naive sense of the term.[9]

A further specification of the phenomenological meaning of the concepts of immanence and transcendence leads necessarily to the doctrine of intentionality. Phenomenology is concerned with the essence of knowledge as revealed in direct 'seeing'. However, this does not mean that it must restrict itself to the genuinely immanent; on the contrary, its major concern is precisely with the immanent in its intentional character. Cognitive phenomena according to their very essence have an *intentio*, they refer to something, they are related to objects. In other words what phenomenology wishes to bring to light are not just mere immanent phenomena, but phenomena which, although immanent, necessarily refer to something which is neither a cognitive phenomenon, nor genuinely contained within a cognitive phenomenon.[10]

One might have the impression from what has just been said that the task which phenomenology assigns to itself is an easy one: it wishes merely to limit itself to a careful description of conscious phenomena insofar as their essential structure can be brought to the fore with apodictic evidence. But Husserl argues that in reality it is not at all easy to achieve such a task. At first sight evidence seems to be only a question of seeing; my intuiting grasp sees the things which are simply there before me; they are there in consciousness in a genuinely evident intuition; my intuiting glance simply looks at them, and takes them just as they are in themselves.

Upon closer investigation and analysis, however, this phenomenological intuition appears to be completely different from a simple act of seeing. Within the domain of phenomenology it is meaningless to speak of 'things which are simply there in consciousness and only are to be intuited'. The 'simply being there of objects' appears to imply a complex of different experiences possessing their own specific structures, such as perception, imagination, memory, retention, and expectation. Things are not in consciousness as are matches in a box. But in and through our experiences things are constituted in such a way that they are in no way really contained in these experiences. For everything 'to be given' means to come into being in this or that way in and through our lived experiences and to be presented by them. Moreover, things are not there first and then impress 'pictures' of themselves upon our consciousness. In the phenomenological reduction the things are 'in' the appearances, and only because they appear to us are they given in themselves. It is true that they appear as being independent of this individual appearance, but essentially they are inseparable from the appearing itself.

This amazing intentional relationship between the knowing act and the known object is found everywhere. The task of phenomenology is not just to intuit this relationship and to describe it, but to explain how an object of knowledge is constituted in our knowing acts. Within the sphere of pure self-givenness and apodictic evidence we must analyze and describe all the modes of presentation and the correlates which become manifest in them. Only in this way shall we see how the transcendent objects are known in our acts of knowledge, namely as that which at the outset was 'meant' only and, also, how this intended meaning is continuously fulfilled in the progressing coherence of knowledge in original acts of synthesis. Then we shall experience how the objects of knowledge are continuously constituted, and how in each case this constitution is already predelineated by the fact that these objects essentially require such a gradual constitution. Phenomenology studies in a genuinely scientific and radical way the constitution of the various objects of consciousness and also the world in which they appear and the ego to whom they manifest themselves.[11]

Husserl concludes this description of his phenomenological philosophy by again pointing to the fact that all of this is to be investigated in the sphere of pure evidence in order to shed light on the great problems of knowledge, that is on the problems of the nature of knowledge and the meaning of the correlation of knowledge and its objects. At first the problem to be examined seemed to be the relation between subjective psychological experiences and that which is actually grasped in them as it is in itself. Now it becomes clear that the crucial problem has to do with the relation between *knowledge* and its *object*, and this in the reduced sense: we are dealing here not with concrete knowledge, but with knowledge in general, independent of any existential assumption concerning the empirical ego and the real world. Furthermore, we must realize again that this problem of intentionality and constitution can be solved only within the sphere of pure evidence, that is within the sphere of data which are ultimate norms because they are absolutely given. And finally one must keep in mind that phenomenology must examine all the fundamental forms of knowledge with their correlative objects one by one by the process of intuitive seeing in order to determine the meaning of all the correlations which are to be explained.[12]

As far as I can determine, Husserl never changed his view on this basic conception of phenomenology in any essential way after 1907, although it is true that he extended his phenomenology from cognitive phenomena to all conscious phenomena, and that he made repeated attempts to determine as accurately as possible the final meaning of this conception, and to found it in a completely radical way. If we do not pay too much attention to the verbal formulation and keep in mind the gradual development of Husserl's thought between 1887 and 1907, his conception of phenomenology as formulated in *The Idea of Phenomenology* seems initially to be quite clear. Perhaps one could characterize Husserl's view by saying that he slowly moved from psychologism, as this was quite generally found in the latter part of the 19th century especially among people interested in philosophy of sciences, to a certain form of transcendental idealism, and that in this development the works of Descartes and Kant played important roles.[13]

Although I know of no one who would deny this general characterization of Husserl's conception of philosophy (a characterization which otherwise can be easily justified with explicit remarks of Husserl himself clearly pointing in this direction)[14] yet for some forty years scholars have been unable to agree on the interpretation of the metaphysical position which Husserl's phenomenology necessarily implies. The question of whether his thought should be interpreted *ultimately* in a realistic or idealistic way is difficult to answer in a few short statements. Some interpreters hold that Husserl's phenomenology is realistic in essence; others see his work as some form of idealism, the exact nature of which cannot easily be determined. There is also an existentialist interpretation of Husserl's phenomenology which argues that Husserl transcends the opposition realism – idealism.[15]

Since I have elsewhere indicated my motives for interpreting Husserl's philosophy as a whole in an idealistic way and have carefully delineated in how far this label affects Husserl's philosophy as a whole[16], I wish to limit myself to a consideration of the question of precisely how Husserl's idealist position is to be specified with respect to Kant's conception of transcendental idealism.

In connection with Husserl's idealism it has often been pointed out that for Husserl Being (*Sein*) is identical to being object, that for him phenomenon means Being, and that intentionality in his view is merely a cognitive and certainly not an ontological relationship. It has also been said that Husserl's view of reduction and constitution necessarily leads to transcendental idealism. However important these remarks may be (as I have attempted to show elsewhere)[17], I nonetheless believe that they do not go to the heart of the matter. Therefore, in trying here to answer this complicated question concerning Husserl's metaphysical position I shall take my starting point in an examination of Husserl's conception of world-constitution which in my view more than any other theme mentioned is the one which most clearly illuminates Husserl's final position on the question concerning realism or idealism. The choice of this topic has the further positive aspect of permitting us to explicitly compare Husserl's phenomenology with Heidegger's conception of philosophy whose quite clear view on the issue will help us, I feel, to shed some new light upon the other elements brought up in the long discussion on the subject.

II. TRANSCENDENTAL SUBJECTIVITY AND THE CONSTITUTION OF THE WORLD

In the first chapter of a remarkable book on Husserl's philosophy Diemer pointed out that Husserl tried to characterize his phenomenology several times by saying that the basic concern and aim of phenomenology consists in an attempt to answer the question concerning "the meaning (*Sinn*) and Being (*Sein*) of being (*Seiendes*)"[18]. One could remark here that this characterization in no way distinguishes phenomenology from any other conception of philosophy in that many philosophers since Aristotle have claimed that the basic concern of philosophy is to be found in considering "being as being" and, thus, in the question concerning the "Being of being". One could go still further and say that it is even the task of the sciences to answer the question which Husserl ascribes to his phenomenology; does physics not try to explain nature, does biology not deal with living things, psychology with man, theology with God, and do not all of them try to explain the "meaning and Being" of the beings mentioned? It is obvious that in making this claim Husserl had something special in mind which, indeed, separates his philosophy from classical philosophy as well as from science.[19] We can elaborate this further following Diemer's ideas on the subject.

As far as philosophy is concerned, classical philosophy since Aristotle's time has taken its starting point in "phenomena" and tried from there to penetrate into the "*archai*" which found these phenomena. Classical philosophy has tried to materialize this ideal by attempting to find a way from the given phenomena to the essences or principles which ultimately found them, or by radically explaining the world (understood as *aggregatum rerum finitaru m*as Leibniz calls it) by pointing to an absolute ground which is God. But in this way classical philosophy tried to go from the one being to another claiming that the latter is the ultimate and, therefore, unquestionable root of the former's appearances. It does not penetrate the question concerning the ground of the Being of being, that is as Husserl formulates it, the question concerning the constitution of being's Being.

Expressed another way, when Husserl says that phenomenology's main concern consists in the question concerning the "meaning and Being of being", he is not primarily interested in the ultimate cause of finite beings, but in answering the question of *how* the Being of being and *how* the Being of the world is to be determined, *how* Being becomes constituted. The ultimate "point" phenomenology comes to in its investigations is thus not an absolute being which creates the world (God), but it tries in questioning to go backwards from the phenomena, that is from the world, to that particular being which questions their Being and to determine the basic structure through which the phenomena manifest themselves in their Being character. The phenomenological absolute is no longer God who creates the world, but the human subjectivity which does not create beings, but merely founds their Being by constituting it.[20]

From this it becomes clear that phenomenology's methods must consist in an 'unnatural return'[21]: the phenomenological method is not oriented toward the world, the mundane things, but goes from mundane phenomena back to the "conditions of the possibility of world experience". But if it is this that phenomenology wishes to accomplish, then the meaning of the reduction becomes immediately understandable. Its meaning is not to deny the world, nor to leave it out of consideration altogether, but merely to show that Being cannot be without subjectivity just as subjectivity cannot be without being, that is (in the final analysis) without the world whose Being the subjectivity constitutes.[22]

As for the sciences, it is obvious in Husserl's view that they are "naive" in that they remain within the realm of the natural attitude. They never found any phenomena *ultimately*, but merely lead us to scientific 'concepts' forgetting that these concepts are not genuinely original. Scientific concepts are 'secondary' constructs which originate from a primary and pre-theoretical source which is our life-world. And what is true for science is analogously true for art, law, state, religious institutions, etc.; all of these 'ideal' constructs originate from the primordial life-world. Phenomenology's first task, therefore, is to 'return' from the ideal constructs to the life-world in order to first explain this world in its typical character and basic structures.[23]

Husserl claims however that even this fundament is not an ultimate one, if one really hopes to explain the constitution of Being. For the life-world is determined by our experience of world.[24] Experience now understood as referring to any primordial encounter of the subjectivity with being in its totality, is primarily achieving activity (*Leistung*) in and through which experienced Being is present to the experiencing subjectivity.[25] However, if experience is primarily achieving activity of the subject, then one has to include this subjectivity in his investigations. But this means that each form and each structure of the life-world requires a return to this subjectivity, that is to say, within the context of a phenomenological inquiry, to the subjectivity of the philosopher who performs the investigation because only there does the philosopher find apodictic evidence and only there is the Being of being immediately present to him as 'it itself'.[26] It is in this primordial ego (*Ur-ego*) as Husserl calls it, that the philosopher finds an absolute point[27] in regard to which everything else is relative.[28]

Phenomenology thus begins with a radical "egology", a "monadology";[29] only after the constitution of the transcendental intersubjectivity is justified will it be possible to turn to an explanation of the "meaning and Being" of our 'objective world valid for everybody', history, language, institutions, etc. All of this evidently does not mean that Husserl's phenomenology is a form of solipsism in the usual sense of the term, although Husserl sometimes speaks of a 'solipsistic egology'.[30] What is usually called solipsism is in Husserl's philosophy transcended the moment it is able to show that consciousness is essentially intentional and, thus, necessarily transcends itself. For then it becomes clear that just as the Being of being cannot be present without the subjectivity, so in the same way the subjectivity cannot be what it is without that being whose Being it constitutes. However, and this is a point of the greatest importance, for Husserl the primordial ego (*Ur-ego*) is essentially a pre-mundane being and in that sense a 'transcendental subjectivity'.[31] It is important to explain here in what sense Husserl understands this.

We have seen that for Husserl the basic concern of phenomenology consists in the question concerning the meaning and Being of being. In making this claim Husserl pre-supposes that Being is identical with 'intelligibility', and thus that Being can and must be reduced to Reason.[32] In other words phenomenology not only claims that knowledge is constituted by consciousness, but also that the very Being of that which is known is so constituted also. For, according to Husserl, only 'absolute Being' is Being in the full sense, and only consciousness and Being-in-consciousness are forms of absolute Being which, therefore, necessarily belong together. Since Being is in consciousness only as constituted by consciousness, Being is absolute only as constituted by consciousness. This is the reason why intentional constitution becomes a universal clarification of Being. Nothing really *is*, except by a proper achieving activity of consciousness. If, then, the task of phenomenology is to understand Being radically, its method must be to penetrate that subjectivity wherein Being has its last source and to clarify the original constitution of Being in and through transcendental subjectivity.[33]

In concretely illustrating these claims Husserl sometimes takes his starting point in an analysis clarifying the constitution of the 'meaning and Being' of a perceptible body with its inner and outer horizon. The outer horizon points to the necessity of carefully examining the constitution of the 'meaning and Being' of the world, in order, then, to proceed to the clarification of the constitution of the 'meaning and Being' of the most important mundane regions.[34] Here we may, following Husserl, take our point of departure in the perception of a die. According to Husserl, constitutive analyses show that this die is given continuously as an objective unity in a multiform and continually changing multiplicity of manners of appearing. In their temporal flow these acts of perception are not an incoherent sequence of lived experiences, but rather merge in the unity of a synthesis in such a way that in them one and the same die is intended as appearing. The one identical die appears now in close appearance, now in distant appearances, in the changing modes of 'here' and 'there', over against an always co-intended 'absolute here'. Moreover, each manner of appearance in such a mode shows itself to be the synthetic unity pertaining to a multiplicity of manners of appearance belonging to that mode. Thus, the die here near by, appears as the same now from this side, now from that, and in every case shows me a determinate perspective. The same holds true for every aspect and property of the die. We always find the aspect in question as a unity belonging to a passing flow of multiplicities of acts. Thus, each passing act intends its object as an act possessing a describable structure of multiplicities that has a definite noetic-noematic composition which, by virtue of its essential character, belongs to only this identical object. As Husserl sees it, the question is to be asked now of *how* we can explain and describe the fact that we experience a multiplicity of appearances given in a multiplicity of acts, as the perception of one single thing? Husserl tries to answer this question by suggesting that only in clarifying the concept of synthesis does the discovery of consciousness' intentionality really become fruitful.

The fundamental form of synthesis is identification. We first find it as an all-encompassing, passively flowing synthesis in the form of the continuous consciousness of internal time. Every experience has its own internal temporality. In the experience of the die, for instance, we can make a distinction between the objective temporality of the die itself and the internal temporality of our die-perceiving. The act in and through which the die appears, flows away with the temporal phases which are the continually changing appearances of this one and identical die. The unity of these phases is a unity of synthesis. It does not consist in the fact that they are stuck to one another externally, but in the fact that they are connected in such a way that they form a unity, the unity of this particular consciousness in which the unity of one particular intentional object as the same object belonging to the multiple modes of appearance becomes constituted. The one identical die is appearing as continuously *immanent* in the flowing stream of consciousness, it is descriptively *in* it, not as a real component, but as something intentional, that is as the immanent objective meaning of that particular form of consciousness. This object of consciousness, just as identical with itself during the flowing experience which involves a multiplicity of acts, therefore, does not come into the experience from without, but is an intentional 'effect' produced by the synthesis of consciousness itself. In other words, the immanent objective meaning as a whole is also constituted in and by consciousness.[35]

It is obvious that one and the same die can also be intended in and through completely different types of acts, simultaneously or successively in different and separate perceptions, recollections, expectations, a.s.o. But here too, there is a synthesis which gives rise to consciousness of identity of one and the same immanent objective meaning.[36]

Moreover, in every consciousness of a plurality and a relational complex there is a synthesis taking place. And finally the whole of conscious life is unified synthetically: it is an all-embracing *cogito*, synthetically comprising all particular conscious experiences and intending the world as the horizon of every particular horizon. It is the task of the constitutional analyses of phenomenology to analyse and describe the constitution which takes place in all these syntheses.[37]

But even if we limit ourselves to the example of the perception of a die, the process of synthesis is obviously more complex than this brief indication of its main lines suggests at first sight. A thorough phenomenological description of the synthesis which takes place in every constitution of an object, and even in the most simple objects like a die, would require longwinded and often complicated analyses in which many new distinctions are to be introduced and special attention would have to be paid to many new aspects which can only come to light in a more careful analysis of an act of perception.[38]

Since it is impossible to go into details here, I shall limit myself to drawing attention to some very important consequences of Husserl's conception of constitution which immediately follow from the few remarks we made on the subject. There it was said that every particular appearance of the die is never more than a onesided and therefore somehow 'imperfect' appearance. From this it follows necessarily that a thing which belongs to the 'real world' can, within the *finite* limits of appearance, appear only inadequately. And from this it follows, in turn, that no rational positing which is founded on appearances which present themselves necessarily inadequately, can be definitive. In other words there are a great number of objects, particularly all those which are included under the rubric 'Nature' and 'World', which in principle cannot be given with complete determinacy and with an equally complete intuition in any finite limited consciousness.[39] This, however, as Husserl argues, does not alter the fact that we nonetheless find definitely described in the *essential nature* of every category of formative synthesis which is the correlate of every category of the object, the *possible* shapes which it can concretely take. Thus it is prescribed on essential lines for every imperfect synthesis of this kind how it may perfect itself, how its meaning may be completed, intuitionally filled, and how this intuition is to be further enriched. In other words every category and every region of the object is a general essence which on grounds of principle must itself be brought to adequate givenness. And in its adequate givenness it prescribes a transparent general rule for every special object of which we become aware in the variety of our concrete experiences. It prescribes the rule which determinates how an object subordinate to it in regard to its meaning and mode of presentation, is to be brought to adequate primordial givenness. Formulated differently, to every object that truly is, there corresponds intrinsically in the *a priori* of the unconditional generality of its essence the idea of a possible consciousness in which the object itself can be grasped in a primordial and also perfectly adequate way.[40] We may conclude therefore that although it is true that a material thing can only be given inadequately, as Idea in the Kantian sense, its complete givenness is nonetheless prescribed, as a connection of endless processes of continuous appearing, absolutely fixed in its essential type.[41] We must now reformulate this and try to express the main idea more carefully.

We have seen that neither the identity of any object meant nor that of a truly existing object, nor the identity constituted in the adequation of the object meant as such with the object that really is, is genuinely an *intrinsic* moment of the transient lived experience of evidence and verification. But there is an *ideal* immanence referring to further complexes of possible syntheses as complexes that play an essential rôle here. For, every evidence founds for me an abiding possession. I can always return to the reality beheld in itself in a series of new evidences securing the first evidence. For instance, in the case of evidence of immanent data, I can return to them in a series of intuitive recollections which, as a potential horizon, creates the 'I can always do so again'. Without such possibilities there would be for me no fixed and abiding being, no real nor ideal world. For both of these worlds exist for us only thanks to the evidence, or the presumption, that we will always be able to 'make evident' in the future, or to repeat acquired evidences of the past.

Thus, the particular evidence itself of one particular act does not as yet produce for us any abiding being. We must distinguish here between everything that is 'in itself' and the accidental being-for-me of this particular act with its particular noema. On the other hand, however, the in-itself refers us also to a certain evidence, not a particular one as this factual experience with its factual correlate, but to determinate possibilities or potentialities which are grounded in the transcendental ego and its transcendental life, namely: (1) to the infinity of intending acts which all relate us to something as identical, and (2) to potential evidences which, as particular and factual experiences, are repeatable *in infinitum*.[42]

In another and even more complicated manner evidences refer to infinities of evidence relating to the same object. For they make their object itself be given with an essential onesidedness because of the perspectivism and the inner horizon both of which are characteristic for every real and perceptible thing. This onesidedness is found not only in the totality of all evidences of the real and objective world as a whole, but also in every particular object that belongs to it.

The evidence pertaining to particular objects in a real and objective world is rooted in 'external experience'; in this case there is no other mode of self-presentation for such an object. As we have seen, this kind of evidence has an essential onesidedness which entails a multiform horizon of unfulfilled anticipations, which are continually in need of fulfillment. This imperfect evidence can be made more and more perfect by means of the actualizing synthetic transitions from evidence to evidence, but there is no imaginable synthesis here which is ever completed as an adequate evidence, because any such synthesis must always involve unfulfilled meanings. At the same time, there always remains the open possibility that the belief in being will not be fulfilled and that what seemed to be does not exist. External experience verifies objects only as long as the passively and actively continuing experiences together constitute a harmonious synthesis. That the Being of the world is, and necessarily remains, transcendent to consciousness does not alter the fact that it is only by our conscious life that everything that is transcendent becomes constituted as something which is inseparable from consciousness. In other words, only conscious life, as world-consciousness, bears within itself inseparably the meaning 'world', 'this really existing world'. Only an uncovering of the horizon of experience ultimately clarifies the reality and transcendence of the world and, at the same time, shows the world to be inseparable from the transcendental subjectivity which ultimately constitutes its meaning and Being.

The reference to harmonious infinities of further possible experiences manifestly signifies that every real object belonging to the world, as well as the world itself, is an infinite Idea, related to infinities of harmoniously combinable experiences – an Idea that is the correlate of the idea of a perfect experiential evidence and a complete synthesis of possible experiences.[43]

III. HUSSERL'S TRANSCENDENTAL IDEALISM

In attempting to characterize his metaphysical position in regard to other conceptions of philosophy, Husserl often used the expression 'transcendental idealism'.[44] The use of this term places Husserl within the general perspective of Kant and neo-Kantianism, and Husserl is, although perhaps somewhat reluctantly, well aware of this relationship. From the context in which Husserl uses the expression as well as from his phenomenological philosophy as a whole it becomes clear that it was never his intention to identify phenomenology with Kantian philosophy, nor even to state that phenomenology and Kant's critical philosophy completely agree in the basic philosophical question concerning realism or idealism. It seems that Husserl uses the term as a label which refers to a general philosophical climate in which he thinks his philosophy belongs, but the basic conception of which he tries to re-interpret.[45] This claim is to be examined and clarified here in greater detail.

It is a well-known fact that Husserl has been of the opinion that Descartes as well as Kant had great influence on the origin of the idea of his phenomenology. It is known also that whereas Husserl's interest in Descartes gradually decreased, his interest in Kant continuously increased. Not only did Husserl study Kant more and more carefully and extensively, but he also found himself approaching Kant's basic conceptions in many important aspects of his own philosophy. In many instances his study of Kant and neo-Kantian philosophers seems to have been the major source of inspiration as far as particular topics of his phenomenology are concerned. True, Husserl repeated time and again that there is a great difference between Kant's critical philosophy and his phenomenology, but this difference does not so much consist in a different conception of philosophy or in a different metaphysical position in regard to the world, as it does almost exclusively in a difference of approach, method. Describing his own development in letters to Rickert, Natorp, and Cassirer, Husserl makes it quite clear that he has come to his philosophy independent from the Kantian philosophy, but that using different methods and going along different roads he gradually finds himself closer and closer to the metaphysical position defended by German idealism in general and Kant's transcendental idealism in particular.[46] Later he adds to this general impression that his view on constitution and synthesis is deeply influenced by Kant and Natorp as far as its genetic conception of them is concerned. But then, too, he adds immediately that the goal Kant hoped to reach and was never able to materialize, can be realized by his phenomenological philosophy: "transcendental phenomenology is an attempt to materialize the deepest meaning of Kant's philosophizing".[47] "Phenomenology finds itself in harmony with Kant's philosophy as far as the final results it is able to come to, are concerned."[48]

In an important study devoted to a careful comparison of Kant's and Husserl's philosophies Kern comes to the conclusion that the relationship between criticism and phenomenology can perhaps best be described in the following way. First of all one must realize that Husserl's conception of phenomenology after its basic idea was discovered, gradually developed in dialogue with and under the continuous influence of Kantian and neo-Kantian philosophy. In addition to this general influence, one must secondly point to several basic topics and conceptions in Husserl's phenomenology whose origin may also be found in Kantian philosophy. One could mention here, among other topics, Husserl's idea of objective logic, his conception of ontology and of the analytic and synthetic *a priori*, the distinction between two fundamental attitudes, the natural and the philosophical, the doctrine of constitution, synthesis, active and passive genesis, Husserl's final conception of the reduction and, finally, Husserl's idealistic interpretation of world.[49] Kern concludes that all of this notwithstanding Husserl's phenomenology cannot be characterized as neo-Kantianism, mainly because of essential differences concerning philosophy's method. Husserl claims that the methods used by Kant and the neo-Kantians have not been able to materialize the main ideal philosophy tries to achieve; he also claims that this ideal, which he shares with Kant and his successors, has been (or at least can be in principle) materialized by his phenomenology.[50]

We have mentioned several times that there is a close relationship and kinship between Kant's and Husserl's transcendental idealism. In order to be able later to refer to points of agreement and disagreement, we must first try to precisely determine what this expression stands for. We have seen in the foregoing discussion that the process of synthesis in and through which the world becomes constituted for the transcendental subject, is conceived by Husserl as an 'achieving activity' (*Leistung*) or productivity (*Produktivität*). Husserl believed that this conception of synthesis indeed had its origin in Kant's conception of synthesis, but, on the other hand, that it is certainly also re-interpreted from the viewpoint of his genetic phenomenology; Husserl explicitly eliminates, among other things, Kant's doctrine of the 'thing in-itself' and his conception of the function of the 'intellectus archetypus'. Finally Husserl thought that, limiting himself to Kant's theory of the constitution of the phenomenal world through the transcendental functions of transcendental subjectivity, phenomenology and critical philosophy are in agreement and that this basic conception can be characterized as transcendental idealism.[51]

Thus according to Kant and Husserl the experienced real world is a product (*Produkt, Gebilde*) of the constituting subjectivity. Neither Kant nor Husserl interprets this conception in a solipsistic way. Kant avoids solipsism by his doctrine of the 'thing in-itself', whereas Husserl avoids solipsism by claiming that although transcendental subjectivity constitutes the world in and through an intentional genesis, the being which is intended in this achievement is nonetheless not constituted by consciousness. In other words the genetic constitution of what as in-itself is foreign to consciousness, is indeed production, but what is produced here is not the in-itself but the in-itself's being-for-me. Insofar as that which as in-itself is foreign to consciousness, possesses a being-for-itself, it is obviously not the product of my constituting activities, but something which in regard to my consciousness is something absolute. Since now the world does not have any for-itself, but merely is a being-for-us, since, in other words, the Being of the world is its Being-for-us, the world itself is considered from the viewpoint of genetic constitution nothing but a product of the constituting subjectivity.[52]

Kern rightly points out that Husserl's conception of the ideality of the world, therefore, fundamentally rests upon two different grounds: (1) the fact that the Being of the world obviously consists in a Being-for-the-experiencing-subjectivity and, thus, that the world from a static point of view is essentially relative to consciousness[53], whereas (2) from a genetic point of view it is the product of the achieving activities of transcendental subjectivity[54]; in other words the second ground of Husserl's idealism is the fact that constitution must be conceived of as something which is essentially genetic.[55]

In briefly elaborating these two different points we must first ask the question of why Husserl is convinced that the real world does not have a Being-for-itself, but merely a Being-for-us. There cannot be any doubt that Husserl is unable to justify this claim by pointing to the thesis according to which there is a strict correlation between consciousness and Being, since it does not follow that the fundamental intelligibility of Being entails that the corresponding being does not have any Being-for-itself. In other words, it is not the transcendental reduction which, indeed, shows the basic correlation of Being and consciousness that permits Husserl to conclude to a fundamental relativity of the world. The reason for Husserl's claim is to be found in what we have said above about the experience of real things. There we drew attention to the fact that in Husserl's view the *perception* of things, which is the primordial dator act in his view[56], is necessarily perspectivistic and that the infinitely open perceptual process is subject to the condition that appearances to be actualized at later phases of the process will agree and conform both with one another and with those appearances actualized at earlier stages of the process. Since the fact that this condition has been fulfilled in the past with respect to a certain material thing does not lead to any unchangeable certainty that this condition will be fulfilled throughout the future course of the perceptual process as a whole, Husserl concludes therefore, that, what we call this material thing is no more than an idea in the Kantian sense. The material thing is no more than the very idea of an infinite system of appearances which can be realized in actual experiences.[57] Husserl furthermore believes that what is true for this particular thing is obviously true for the world at large.[58]

In order to avoid all possible misunderstandings we must explicitly remark here that it is not Husserl's intention to deny the reality of the world.[59] Husserl has repeatedly warned of this possible misunderstanding.[60] Anyone who attacks Husserl's view on this ground manifests that he does not understand the meaning of the reduction and still is in the realm of the 'naive' and natural attitude. Husserl is not trying to reduce one being (this chair, the totality of things) to another being (God for instance, or to transcendental subjectivity conceived of as mundane being); it is not his intention in other words to give a foundation to beings, but he is trying to found their 'Being and meaning'.[61] Berger has formulated Husserl's view in this regard with great clarity: "Phenomenological idealism does not deny the true existence of the world. The only task consists in elucidating the meaning (= Being) of this world. There is no doubt that the world exists, but one has to try to understand this indubitability".[62]

In trying to establish this indubitability Husserl is of the opinion that in contradistinction to the real world, transcendental subjectivity is something absolute. "Only subjectivity can be 'for-itself' in a genuine and absolute way. For Being-for-itself is appearing-to-itself ..."[63] "Absolute being is Being in the form of an intentional life which, whatever it may be concretely conscious of, is at the same time consciousness of itself."[64] Transcendental subjectivity does not depend upon a process in which meaning is given to it (this is characteristic of the world) and which is to be given by another subjectivity.[65]

But all of this does not yet necessarily mean that in Husserl's view transcendental subjectivity is the only and last ground of the world's Being.[66] In order to explain this it is necessary to compare more carefully Kant's and Husserl's conceptions of the relationship between subjectivity and world.

It is a well known fact that Husserl's conception of the ego underwent a long and perhaps even never-ending development. Generally speaking one may say that Husserl originally thought that only one ego is to be admitted: an empirical, bodily-spiritual ego which, from a phenomenological point of view, manifests itself as the complex of all phenomena.[67] At that time he explicitly rejected Natorp's conception of the ego as the pole and reference center of all contents of consciousness, whereas Kant's transcendental ego, the ego of transcendental apperception was not mentioned at all. Between 1900 and 1916 Husserl first adds a pure ego as the individual ego-pole of all acts of consciousness[68] and, then, the ego which in its individuality is constituted by its 'habitualities'.[69] Finally, in 1916, Husserl developed this idea further in that the ego and particularly its identity are not so much connected with the ego's habitual convictions concerning concrete contents of consciousness, but are mainly conceived of as the consequence of the positing (*Setzung*) of the world as universal horizon.[70] It is known also that study of Natorp's works and Kant's conceptions deeply influenced this development in Husserl's thought.[71]

Husserl thus finally came to the conclusion that the pure ego is constituted in such a way that it can maintain itself as this identical ego merely when it is able to maintain the objectivity-thought-of as remaining identical with itself in all its thoughts. In other words, the ego constitutes its own concrete identity by means of its harmonizing and continuously maintained positions in regard to the world. The convictions of the ego in regard to concrete mundane objects may obviously change, but the identity of the ego requires that in all these changes it is determined by the idea of coherence and harmony in all its convictions, which as idea is founded in the always remaining conviction concerning the harmonious world itself.[72]

In summary, Husserl gradually becomes aware of the necessity of admitting that the constitution of the world as an harmonious kosmos (*einstimmiger Kosmos*) is a necessary condition for the unity and identity of the pure ego: the self-identity of the ego requires that notwithstanding all changes of opinion in regard to concrete objects, the ego maintains its basic conviction concerning the harmonious world.[73] This ego, however, is not a pure ego-pole of ego-acts, but an ego which is convinced of its own transcendence in regard to the acts as well as in regard to their objects.[74]

Although Husserl thus gradually developed in the direction of Kant's conception of the ego, there are nonetheless many important differences between these views. The most important of these differences, to which we must turn our attention here, is that in Kant's conception the original and *necessary* consciousness of the ego's self-identity is, at the same time, a consciousness of an *equally necessary* unity of syntheses of all appearances *according to concepts*, that is the ground of the unity of all objectivity. Husserl on the other hand is convinced that the ego of the transcendental apperception which constitutes itself through its constitution of the world, just as well as that which it so constitutes, is no more than a *mere factum*, the moment it is considered from the viewpoint of transcendental subjectivity. It is and remains always possible that the world becomes dissolved in a mere "Gewühl", which obviously entails a dissolution of the ego.[75]

Husserl's relatively complicated position concerning the relationship between ego and world seems thus to involve the following seemingly conflicting theses:

(1) The constitution of the world by the transcendental subjectivity in transcendental consciousness is a mere contingent fact.

(2) The contingence of this fact not only relates to the fact *that* there is a world for consciousness, but also to the *manner* in which the world is for consciousness.

(3) It is nonetheless possible to develop an ontology of the world which describes the necessary and essential structures that determine the world's *So-Sein.*

(4) The contingence of the constitution of the world does not put the world on the same level as the contingent mundane things, in that the world (regardless of my fluctuating opinions about concrete mundane things) is and remains the ideal and always presumed horizon on the basis of which conflicting opinions about concrete mundane things can be brought into harmony with one another.[76]

This view is obviously different from the views of Kant, Fichte, and Natorp. And Husserl was deeply aware of these differences. Husserl agrees with German idealism that the transcendental subjectivity productively brings about the world in and through its own life in genetic constitutions. However, Husserl claims, this constitution does not take place in such a way that it necessarily follows from the very essence and through the sole power of transcendental subjectivity. That the subjectivity can and does constitute *such a world* as it in fact constitutes, is not because of its own creative possibilities, but depends upon a certain 'gift' or 'grace'; the subjectivity can withdraw from this 'grace' and thus the world can always dissolve into a mere chaos of impressions. In other words, the subjectivity, indeed, is the ground of the Being of the world, but it is not the *sufficient* ground of its Being; the constitution of the world is not genuinely the work of the finite subjectivity. It seems that Husserl has taken this insight as the starting point for a 'metaphysics' in the Kantian sense, explaining the constitution of the world as the 'product' of a real and genuine Absolute, God.[77]

If in concluding these considerations we now compare Husserl's and Heidegger's philosophies, then it becomes clear, it seems to me, that Husserl and Heidegger certainly agree on this fundamental idea that the basic concern and aim of philosophy consists in an attempt to answer the question concerning the 'meaning and Being of being'.[78] Both agree, also, that in trying to materialize this goal the philosopher should not be primarily interested in the ultimate cause of all finite beings, but in answering the question of *how* the Being of beings and *how* the Being of the world are to be constituted.[79] Finally both agree that in answering the question concerning the meaning and Being of being a privileged position is to be attributed to the subjectivity, that is, that being which questions the Being of beings.[80] "We agree that being taken in the sense of what you call 'world' cannot be clarified in its transcendental constitution by means of a return to a being of the same kind."[81] The disagreement between these two philosophers is apparent in their completely different views of the transcendental reduction. Heidegger denies the necessity and even the possibility of such a reduction, whereas in Husserl's view there can be no philosophy without such a reduction. Referring to Heidegger's reinterpretation of phenomenology Husserl writes[82]: "... one misinterprets my phenomenology backwards from a level which it was its very purpose to overcome, in other words, that one has failed to understand the fundamental novelty of the phenomenological reduction and hence the progress from mundane subjectivity (i.e., man) to transcendental subjectivity; consequently that one has remained stuck in an anthropology ... which according to my doctrine has not yet reached the genuine philosophical level, and whose interpretation as philosophy means a lapse into 'transcendental anthropologism', that is 'psychologism'." Many other points of disagreement are necessarily connected with this basic difference. In the first part of this essay I have shown how Husserl, at the very start in 1907, was convinced that his phenomenology in principle is able to bring about a philosophy which in every possible sense fulfills all the requirements of a 'strict and rigorous science'. Phenomenology is able to do so because it limits itself to the unquestionable, the apodictic evident, the genuinely immanent; to unseal the realm of the apodictic is the task of the reduction. What has been sealed off by the reduction is then 'brought back' into the subject matter of philosophy by means of phenomenology's conceptions of intentionality and constitution; as the totality of all meaning, the world, however, participates now in all the privileges of the 'genuinely immanent', particularly in the apodictic evidence with which it presents itself to the subjectivity. In all these issues Heidegger fundamentally agrees and equally fundamentally disagrees with Husserl. And if we now ask ourselves what the source of this basic disagreement is, then it is clear, it seems to me, that the root of all philosophical truth (which for Husserl as well as for Heidegger is to be found in transcendental subjectivity) consists for Husserl in a subjectivity which originally is world-less, whereas for Heidegger the 'ultimate' starting point for philosophy is to be found in man as Being-in-the-world. Most revealing in this connection is the fact that both Husserl and Heidegger in speaking about each other's philosophies explicitly mention their conceptions of transcendental subjectivity; but where for Husserl this conception is determined by the demands of his reduction, there Heidegger argues that a genuine understanding of this subjectivity makes such a reduction impossible. "... what is the mode of Being of that being in which 'world' becomes constituted? That is the central problem of *Being and Time*, that is, of a fundamental ontology of *Dasein*. One must try to show that the mode of Being of human *Dasein* is completely different from that of all other beings and that this mode of Being, as that which it is, precisely contains the possibility of the transcendental constitution. Transcendental constitution is a central possibility of the ek-sistence of the *factical* self. This, the *concrete* man, is as such (that is as being) never a 'mundane real fact', since man never is merely present-at-hand, but ek-sists. And the 'marvel' consists in this that the understanding of *Dasein's* ek-sistence makes the transcendental constitution of everything which is positive, possible."[83] In other words, the world obviously is constituted, but in constituting the world man always finds the world already there, because it is a constitutive component of his own Being of which he always has already a preontological understanding.

REFERENCES

1 Edmund Husserl, *Die Idee der Phänomenologie. Fünf Vorlesungen* (1907) (Edited by Walter Biemel), *Husserliana*, Vol. II, Martinus Nijhoff, The Hague, 1950.
2 *Ibid.*, p. 19, pp. 79–80.
3 *Ibid.*, pp. 16–17.
4 *Ibid.*, pp. 17–18.
5 *Ibid.*, pp. 18–19.
6 *Ibid.*, pp. 22–23.
7 *Ibid.*, pp. 33–40.
8 *Ibid.*, pp. 40–42.
9 *Ibid.*, pp. 6–7.
10 *Ibid.*, p. 53.
11 *Ibid.*, pp. 8–12.
12 *Ibid.*, p. 60.
13 Joseph J. Kockelmans, *A First Introduction to Husserl's Phenomenology*, Duquesne University Press, Pittsburgh, Pa., 1967, pp. 1–34.
14 Cf. *Die Idee, Einleitung des Herausgebers*, pp. vii–viii; Iso Kern, *Husserl und Kant. Eine Untersuchung über Husserls Verhältnis zu Kant und zum Neukantianismus*, Martinus Nijhoff, The Hague, 1964, pp. 14–33.

[15] Joseph J. Kockelmans, *op. cit.*, pp. 320–348.
[16] *Ibid.*, pp. 348–355.
[17] *Ibid.*, pp. 351–353.
[18] Alwin Diemer, *Edmund Husserl. Versuch einer systematischen Darstellung seiner Phänomenologie*, Anton Hain, Meisenheim am Glan, 1965, p. 9. Cf. Edmund Husserl, *Formale und transzendentale Logik. Versuch einer Kritik der logischen Vernunft*, Max Niemeyer, Halle a. S., 1929, p. 13, p. 148; *Cartesianische Meditationen und Pariser Vorträge* (ed. by Prof. S. Strasser), *Husserliana*, Vol. I, Martinus Nijhoff, The Hague, 1950, p. 13.
[19] Alwin Diemer, *op. cit.*, p. 9.
[20] *Ibid.*, pp. 9–10. Cf. Eugen Fink, 'Die phänomenologische Philosophie Edmund Husserls in der gegenwärtigen Kritik', *Kant-Studien* **38** (1933) 319–383, pp. 338–339.
[21] Edmund Husserl, *Logische Untersuchungen*, 3 Vols. Max Niemeyer, Halle a. S., 1929–1922, Vol. II/1, p. 9; *Die Krisis der europäischen Wissenschaften und die transzendentale Phänomenologie. Eine Einleitung in die phänomenologische Philosophie* (ed. by Walter Biemel), *Husserliana*, Vol. VI, Martinus Nijhoff, The Hague, 1954, p. 204.
[22] Alwin Diemer, *op. cit.*, pp. 10–12.
[23] *Ibid.*, pp. 12–13.
[24] *F. tr. L.*, p. 206.
[25] *Ibid.*, p. 206.
[26] *Ibid.*, pp. 208–211.
[27] Edmund Husserl, *Ideen zu einer reinen Phänomenologie und phänomenologischen Philosophie*, 3 Vols. Martinus Nijhoff, The Hague, 1950–1952, Vol. I, p. 139.
[28] *Ibid.*, pp. 139–140; *Cart. Med.*, p. 27.
[29] *Cart. Med.*, p. 89, p. 150.
[30] *Ibid.*, p. 154, p. 38.
[31] Alwin Diemer, *op. cit.*, pp. 13–15.
[32] *Ideen*, Vol. I, pp. 329–333; *Cart. Med.*, pp. 91–92.
[33] Q. Lauer, *The Triumph of Subjectivity. An Introduction to Transcendental Phenomenology*, Fordham University Press, New York, 1958, pp. 79–80.
[34] *Ideen*, Vol. I, pp. 313–380; *Cart. Med.*, pp. 91–121.
[35] *Cart. Med.*, pp. 77–81.
[36] *Ibid.*
[37] *F. tr. L.*, pp. 276–280.
[38] Alwin Diemer, *op. cit.*, pp. 115–312 (for the problem of constitution as a whole); Aron Gurwitsch, *The Field of Consciousness*, Duquesne University Press, Pittsburgh, Pa., 1964, pp. 199–305 (for the constitution taking place in an act of perception).
[39] *Ideen*, Vol. I, p. 355.
[40] *Ibid.*, p. 365.
[41] *Ibid.*, p. 366.
[42] *Cart. Med.*, pp. 95–96.
[43] *Ideen*, Vol. I, pp. 348–351; *Cart. Med.*, pp. 97–99.
[44] Edmund Husserl, *Erste Philosophie* (1923–1924), 2 Vols. (ed. by Rudolf Boehm), *Husserliana*, Vols. VII and VIII, Martinus Nijhoff, The Hague, 1956–1959, Vol. II, p. 181; *Cart. Med.*, pp. 174–177, pp. 116–21.
[45] Iso Kern, *op. cit.*, pp. 34–50, pp. 276–303.
[46] *Ibid.*, pp. 28–29.
[47] *Erste Philosophie*, Vol. I, p. 287.
[48] *Ibid.*, p. 235.

49 Iso Kern, *op. cit.*, pp. 421–422.
50 *Ibid.*, pp. 422–423.
51 *Ibid.*, pp. 276–277. Cf. *Cart. Med.*, p. 118, p. 176.
52 *Erste Philos.*, Vol. I, pp. 270–280.
53 *Ideen*, Vol. I, p. 159.
54 *Erste Philos.*, Vol. I, pp. 274–276.
55 Iso Kern, *op. cit.*, pp. 279–280.
56 *Ideen*, Vol. I, p. 11.
57 *Ibid.*, p. 350–351, 365–374.
58 *F.tr.L.*, p. 249; *Erste Philos.*, Vol. II, p. 47. Cf. Iso Kern, *op. cit.* pp. 280–283.
59 *L.U.*, Vol. II/2, pp. 235–236.
60 *Ideen*, Vol. I, pp. 63–69; Vol. III, pp. 152–153.
61 Alwin Diemer, *op cit.*, pp. 16–17.
62 Gaston Berger, *Le cogito dans la philosophie de Husserl*, Aubier, Paris, 1941, p. 95.
63 *Erste Philos.*, Vol. II, p. 189.
64 *F.tr.L.*, p. 141.
65 Iso Kern, *op. cit.*, p. 284.
66 *Ibid.*, p. 285.
67 *L.U.*, Vol. II/1, Section 8.
68 *Ideen*, Vol. I, p. 138.
69 *Ideen*, Vol. II, pp. 111–120; *Cart. Med.*, pp. 100–101.
70 *Erste Philos.*, Vol. I, pp. 395–400.
71 Iso Kern, *op. cit.*, pp. 286–288.
72 *Ms.* orig. A VI 30, p. 34b–44a (quoted by Kern, *op. cit.*, pp. 289–290).
73 Iso Kern, *op. cit.*, pp. 290–291.
74 *Ibid.*, pp. 291–293.
75 *Erste Philos.*, Vol. II, pp. 44–58; *Ideen*, Vol. I, pp. 114–115.
76 Iso Kern, *op. cit.*, pp. 295–296.
77 *Erste Philos.*, Vol. I, pp. 381–395; p. 220; pp. 187–188 (note 1); Vol. II, pp. 254–255; Iso Kern, *op cit.*, pp. 297–303.
78 Martin Heidegger, *Sein und Zeit*, Max Niemeyer, Tübingen, 1963, pp. 2–15.
79 *Ibid.*, pp. 200–212. *Identität und Differenz*, Neske, Pfullingen, 1957, pp. 35–73.
80 *Sein und Zeit*, pp. 15–19.
81 M. Heidegger, *Letter to Husserl*, October 22, 1927 quoted by Biemel in Edmund Husserl, *Phänomenologische Psychologie* (1925) (ed. by Walter Biemel), *Husserliana*, Vol. IX, Martinus Nijhoff, The Hague, 1962, p. 601.
82 *Ideen*, Vol. III, p. 140.
83 *Letter to Husserl*, *op. cit.*, pp. 601–602.

ROMAN INGARDEN†

DIE VIER BEGRIFFE DER TRANSZENDENZ UND DAS PROBLEM DES IDEALISMUS IN HUSSERL

Um eine philosophisch grundlegende Auslegung der Philosophie Edmund Husserls und deren Bewertung der grossen philosophischen Probleme, darunter das Problem Idealismus-Realismus, zu erreichen, muss man sich mit den entscheidenen Begriffen, die Husserl anwendet, auseinandersetzen um ihren präzisen Sinn zu gewinnen. Es handelt sich hier um den Begriff der Transzendenz, der, wie wir es zeigen werden, sich in Husserls Gedanken für das Problem Idealismus-Realismus entscheidend erweist.

Der Begriff der Transzendenz spaltet sich in vier verschiedene Bedeutungen. Darunter kann man einen als ontologisch, die drei anderen als erkenntnistheoretisch bedingt betrachten. Diese letzten drei Bedeutungen der Transzendenz, oder besser drei Begriffe der Transzendenz, sind aufs innigste mit Husserls Auffassung der Wahrnehmung verknüpft. Wir fangen nun mit einer Skizze der Theorie der Wahrnehmung an, dann wird zu dem ontologischen Begriff der Transzendenz der Übergang gebahnt; zuletzt wird die Bedeutung dieser Erörterung für das Problem des Idealismus-Realismus im Husserls Denken erscheinen.*

I

Es gibt mehrere Behauptungen über die Wahrnehmung bei Husserl mit denen man anfangen soll. Die erste lautet: Jede Wahrnehmung und insbesondere jede äussere Wahrnehmung, ist ein – wie Husserl sagt – "thetischer" Akt. "Thetischer Akt", ist ein Akt, in dessen Vollzug die Seinsanerkennung dessen, was wahrgenommen wird, vollzogen wird.

Zweitens: jede äussere Wahrnehmung und jede Wahrnehmung überhaupt, ist ein Akt, in dem etwas selbst präsentiert wird; nicht "repräsentiert", sondern selbst präsentiert, selbstgegenwärtig wird, als selbstgegenwärtig gegeben, und zwar in einer ganz besonderen Art der Anschaulichkeit, die mit gar keiner anderen Anschaulichkeit verwandt ist.

Drittens: jede äussere Wahrnehmung, insbesondere jeder Wahrnehmungsakt, ist Glied einer Mannigfaltigkeit von Akten, die inhaltlich

Tymieniecka (ed.), Analecta Husserliana, Vol. I, 36–74. *All Rights Reserved*
Copyright © 1970 *by D. Reidel Publishing Company, Dordrecht-Holland*

zueinander gehören. Sie beziehen sich alle, soweit sie diese Mannigfaltigkeit bilden, auf ein bestimmtes Ding oder eventuell eine bestimmte gegenständliche Situation, oder auf einen bestimmten Vorgang. Dass sie zueinander gehören, dass also die einzelne Wahrnehmung in diesem Sinne inhaltlich nicht ganz selbständig oder von den anderen Wahrnehmungen unabhängig ist, dies zeigt sich darin, dass jede aktuelle Wahrnehmung durch andere, frühere Wahrnehmungen auf verschiedene Weise motiviert wird. (Die Bedeutung dieses "Motivierens" bildet eine ganz besondere Frage für sich. Das Problem der Motivation kann hier aber nicht entwickelt werden.) Jede zu einer solchen Mannigfaltigkeit gehörende äussere Wahrnehmung weist auch auf die nachfolgenden Wahrnehmungen hin und motiviert sie in ihrer Weise. In dem ganzen Verlauf einer solchen Mannigfaltigkeit werden die jetzigen Wahrnehmungen durch spätere Wahrnehmungen entweder bestätigt, bekräftigt oder entkräftigt. Es tritt dann s.z.s. eine gewisse, unerwartete Lösung dessen ein, was jetzt gegeben und angekündigt wird.

Viertens: jeder Wahrnehmungsakt ist noch auf eine andere Weise in gewissem Sinne strukturell unselbständig. Er ist nämlich umrahmt einerseits durch die sogenannte Retention, die sich auf das bezieht, was soeben vorbeigegangen ist, oder gerade vorbeigeht; und andererseits durch eine Protention, die sich auf das bezieht, was gerade eintritt, nachfolgt, was zum Erscheinen kommt. Beides, und darin vor allem die Retention, ist eine unentbehrliche Bedingung dafür dass einem in der Wahrnehmung das weiterhin identische Ding gegeben wird, dass die Identität nicht zerreisst, dass man bei demselben Ding verbleibt. Wenn diese Identität aufhört zu bestehen, dann steht man vor einer unerwarteten Situation und infolgedessen tritt zunächst eine gewisse Desorientierung ein.

Fünftens: jede Wahrnehmung, jede äussere Wahrnehmung, ist, wie ich das nannte, "partiell". Das heisst, wie Husserl sagt, "inadäquat". Das soll bedeuten: selbst da werden nicht alle Bestimmtheiten des wahrgenommenen Dinges in der betreffenden Wahrnehmung anschaulich gegeben. Und gewöhnlich geschieht es so, dass in den einzelnen Wahrnehmungen immer andere Seiten des betreffenden Dinges zur Gegebenheit gelangen, während die früher gegebenen zum Teil aus den Gesichtsfeld verschwinden. Das ist kein Zufall. So ist es notwendig: das Ding kann nicht zugleich allseitig in allen seinen Bestimmtheiten erfasst, gegeben werden. Und da stossen wir auf *den ersten Begriff der "Transzendenz"*;

das Ding, das wahrgenommene Ding, "transzendiert" die betreffende Wahrnehmung, d.h. seine Bestimmtheiten gehen über das jeweilig Gegebene hinaus.

Sechstens: auch im Rahmen dessen, was von dem Ding gerade gegeben wird, was in den Feld des Gegebenen sich vorfindet, ist ein Unterschied zu machen in der Weise, wie es gegeben wird, und zwar: manche Bestimmtheiten des Dinges werden in einer Mannigfaltigkeit von "erfüllten Qualitäten" gegeben und werden dadurch, im engeren Sinne, "effektiv gegeben"; die anderen Bestimmtheiten, die noch gegeben sind, werden bloss "mitgegeben", "mitvermeint", und zwar nicht in erfüllten, sondern in gewissem Sinne in besonderen anschaulich intentionalen Momenten, die über das Erfüllte so hinausgehen, dass man das in ihnen Bestimmte als selbstgegeben fasst. Da ist etwas Phänomenales an dem erscheinenden Ding, aber qualitativ nicht voll erfüllt.

Und da kann man im zweiten Sinne sagen: die äussere Wahrnehmung, jede einzelne Wahrnehmung, die zu der betreffender Mannigfaltigkeit von Wahrnehmungen gehört, "transzendiert" dasjenige, was auf ganz "erfüllte" Weise zur Gegebenheit gelangt. Es gibt hier also zwei verschiedene Transzendenzen, die beide vom erkenntnistheoretischen Standpunkt als Transzendenzen gelten können. Wenn ich z.B. in einer visuellen Wahrnehmung ein Ding sehe, dann ist das Gesehene freilich keine Mannigfaltigkeit rein visueller Qualitäten, denn es werden nicht bloss Farben und die Gestalt des Dinges gesehen. Wie W. Schapp gezeigt hat, sieht man auch die Glätte oder die Rauhigkeit der Oberfläche des Dinges, so wie z.B. die Materialbestimmtheit, oder dass das eine Ding aus Holz und das andere Tuch ist. Jedenfalls aber gibt es noch weitere Bestimmtheiten des Dinges, wie z.B. seine akustische Bestimmtheiten, die gar nicht wahrgenommen werden. Sie sind natürlich mit dem Ding selbst da, aber dieses transzendiert mit ihnen das Feld, den Bestand der Bestimmtheiten, die gerade wahrgenommen werden. So ist es natürlich in der jetzt betrachteten visuellen Wahrnehmung.

In jeder anderen Wahrnehmung, z.B. beim Hören, Tasten, Schmecken usw. kann das ganz anders sein. Es handelt sich um das Wandelnde, Veränderliche. Jedenfalls aber bleibt immer das in der Wahrnehmung nicht gegebene, das über diese betreffende Wahrnehmung hinausgeht. Vom Standpunkt des Dinges ist es gegenüber der Wahrnehmung transzendent. Es ist mit vielen Eigenschaften erfüllt, es ist reicher als das, was gerade wahrnehmungsmässig gegeben wird. Innerhalb dessen aber, was gegeben wird, wird wiederum getrennt: nämlich zwischen dem in effektiven, erfüllten Qualitäten gegebenen und dem das noch übrigbleibt. Dies ist z.B. das Hintere, die Hinterseite, die doch als Hinteres mitvermeint und mitgegeben wird. Das Ding, diese Bank, dieser Tisch, ist keine "Kulisse", es ist ein volles Ding. Weiter ist es das Innere. Diese hintere Seite kann auch falsch mitgegeben sein. Ich blicke jetzt auf einen Kollegen, der im Saale sitzt, sehe gewissermassen auch seine von mir gerade abgewandte Seite und ich glaube, dass er über seinen Schultern auch eine graue Jacke trägt. Sie ist vielleicht tatsächlich an den Schultern gerade nicht so grau, wie Vorne. Aber es geht von dem, was ich gerade sehe, eine anschauliche Intention darauf hin, was da möglicherweise noch ist, was auch gegeben, auch mitgegeben wird, auf das phänomenal anwesende. Wir sehen doch den betreffenden Mann nicht ohne Rückenteil, nicht mit einem grossen Loch hinten in der Jacke, sondern mit dem Rückenteil und mit einer Jacke, die an den Schultern ebenso "grau" ist wie vorne.

Die Wahrnehmung erhebt sozusagen den Anspruch, das *Ganze* des Dinges gegeben zu haben. Sie sagt gewissermassen mehr über das betreffende Ding, als sie verantworten kann. Denn "verantworten" kann sie bloss dies, was in erfüllten Qualitäten an der Vorderseite des Dinges dasteht, und das Hintere des Inneren ist schon bloss mitgemeint, bloss mitgegeben. Insofern geht also die Wahrnehmung über das echt, im engeren Sinne erfüllt Gegebene hinaus; sie transzendiert das gerade so Gegebene. Wir haben es also mit *zwei verschiedenen Begriffen der "Transzendenz" der Wahrnehmung* zu tun und diese beiden sind erkenntnistheoretisch bestimmt.

Es gibt aber noch etwas, was hier erwähnt werden muss, und was für Husserl sehr bedeutend war. Der kritische Realismus und die Physik, die ganze heutige Physik, auch die ganze Biologie, die Naturwissenschaft überhaupt wird in einem bestimmten Geiste, in einer bestimmten erkenntnistheoretischen Auffassung geschrieben bzw. gedeutet. Es wird nämlich behauptet, dass alles, was gegeben wird und was als gegeben mitvermeint wird und für uns noch da ist, und auch das noch was als Transzendent gilt, das alles wäre "subjektiv". Das ist alles die qualitativ bestimmte Welt und dahinter erst steckt die "wirkliche" materielle Realität. Wie diese Realität verstanden wird, hängt davon ab, in welcher geschichtlicher Epoche wir sie betreten. Zuerst glaubte man z.B. noch nicht an die Existenz der Atome, und meinte, dass Atome eine blosse wissenschaftliche Konstruktion seien, die irgendwie aus solchen oder anderen Gründen in die Wissenschaft eingeführt werden. Bohr meinte dagegen dass es die Atome als eine Realität gäbe und dass sie eine ganz bestimmte Struktur aufweisen. Es sei im Atom ein Kern vorhanden, der von einer Reihe von Elektronen umkreist wird, die sich auf solchen oder anderen Bahnen bewegen. Das war für Bohr die Realität. Später wurde der Atomkern zerschlagen. Und jetzt gibt es, ungefähr 30 oder mehr, "Elementarteilchen", die sich auf verschiedene Weise bewegen und die eine längere oder kürzere Lebensdauer haben. Die letzte Realität ist somit heute diese Mannigfaltigkeit im leeren Raum sich bewegender Elementarteilchen. Sie gilt für das "Wirkliche" und das, was wahrgenommen wird, ist "subjektiv" nicht wahr.

Es besteht also das Problem des sogenannten "physikalischen Dinges". Husserl musste dieses Problem natürlich von seinem Standpunkt aus lösen. Er behauptet dass diese physikalische "Wirklichkeit" eine neue *Transzendenz* sei. Sie ist eine erweiterte Transzendenz. Das physikalische Ding, das heisst die Mannigfaltigkeit von Elementarteilchen, ist prinzipiell nicht nur nicht wahrnehmbar, sondern auf gar keine Weise anschaulich vorstellbar. Dies ist die These der heutigen Physik. Das "Transzendente" in diesem Sinne ist bloss durch mathematische Formeln bestimmt; anders kann es nicht bestimmt werden. Dieses so "Transzendente", ist im Sinne der Physik die Wirklichkeit, während die gesammte wahrgenommene qualitativ bestimmte Welt nur "subjektiv" sein soll. (Wenigstens wird es in der wissenschaftlichen Praxis der Naturwissenschaft so behandelt.) Unter den philosophierenden (theoretischen) Physikern des XX. Jahrhunderts ist die Ansicht darüber geteilt. Es gibt "realistisch" eingestellte Physiker (etwa Max Planck), welche diese Auffassung vertreten. Es gibt aber – wenn man so sagen darf – eine "idealistische" Ausdeutung der physikalischen Grundgegenständlichkeiten (z.B. von Eddington) nach welcher dieselben bloss für begriffliche Konstruktionen gehalten werden. Wenn Husserl dieses annehmen müsste, dann wäre es mit seiner ganzen Theorie, die zum transzendentalen Idealismus führt, aus. Er behauptet dass das Ding natürlich etwas ist, was er als ein höheres intentionales Korrelat von verschiedenen Wahrnehmungen berücksichtigt, die mit Hilfe verschiedener Gedankenoperationen interpretiert werden. Das Niveau der anschaulichen Dinge ist ein sehr hohes

Niveau der Konstitution. Das Physikalische ist darin ein noch höheres Niveau, eine weiter gebrachte Höhe der Abstraktion der Synthese der reinen Gedankenkorrelate, die gar keine direkte Erfüllung in den Gegebenheiten der äusseren Wahrnehmung haben können. Es gibt aber Beziehungen, die zwischen dem einen Niveau oder dem anderen Niveau auf gedanklichem Wege und auf eine bestimmte Weise gestaltet oder entdeckt werden. Das Physikalische ist nicht eine letzte materiale Realität, die durch kausal geregelte Wirkungen auf den menschlichen Körper zu der "subjektiven" Ansicht der qualitativ bestimmten Dinge führt, sondern es ist bloss eine Phase; eine weitere Phase der Konstitution der realen Welt. Merkwürdiger Weise soll das Physikalische eine solche Gegenständlichkeit sein, die unter dieses qualitativ bestimmte anschaulich gegebene Ding untergeschoben wird. Das transzendente materielle Ding ist zunächst so bestimmt, wie es in der Wahrnehmung gegeben wird, und so wird es im täglichen Leben behandelt. Aber dann, wenn man die Ergebnisse der Physik (und Physiko-Chemie) akzeptiert, werden sie mit Hilfe der mathematischen Physik bestimmt. Die Transzendenz der physikalischen Dinge gewinnt damit den Charakter einer erkenntnistheoretischen Auffassung, denn es handelt sich um eine Transzendenz in Bezug auf die Wahrnehmung.

Wir werden noch dazu kommen, wenn wir mehrere andere Begriffe der Transzendenz mit diesem Begriff vergleichen. Sie sind alle sehr wichtig, weil sie später bei der Darstellung der Wege, die Husserl zum Idealismus geführt haben, verwendet werden. Jetzt geht es darum noch weitere Fortschritte in der Analyse der äusseren Wahrnehmung zu machen.

Der nächste Schritt ist wiederum eine These, die man ganz kurz fassen kann: "Jede äussere Wahrnehmung, in der etwas gegeben ist, ist eine Hervorhebung des Gegebenen aus einem grossen Feld, dem Wahrnehmungsfeld, in dem es gegeben wird". Also, wenn ich jetzt z.B. mein gegenüber ansehe, den Herrn Kollegen soundso, dann habe ich ihn wahrgenommen, seine Brille z.B., das heisst: Dieser Herr soundso wird aus dem ganzen Feld der Wahrnehmung, die ich vollziehe, herausgehoben. Zu dem wahrgenommenen Ding gibt es ein Feld anderer Dinge und Vorgänge. Und dieses Feld zieht sich immer weiter zu einem Horizont, wie Husserl sagt, aus, und in diesem Feld gibt es sehr verschiedene Formen der Gegebenheit. Einerseits das, was wahrnehmungsmässig hervorgehoben wird, und andererseits das, was sich noch im Feld befindet.

Husserl führt zwei Begriffe ein, die gewöhnlich etwas falsch verstanden werden. Er spricht von der "Aktualität" und der "Inaktualität" des Bewusstseins, als von zwei verschiedenen Formen des Bewusstseins. Das, was im Akt der Wahrnehmung vor sich geht und zur Erfassung eines Dinges führt, heisst "aktuelles Bewusstsein". Wir würden besser sagen, ein "Aktbewusstsein", ein im Akt bestehendes, sich vollziehendes Bewusstsein. Und die Weise, vermöge welcher uns das ganze Feld noch irgendwie anwesend ist, obwohl wir uns zu ihm gar nicht hinwenden, heisst "Inaktualität", inaktuelles Bewusstsein. Anstatt dessen werden wir den Ausdruck "Non-Akt-Bewusstsein" anwenden. Dieses Non-Akt-Bewusstsein bzw. dessen Korrelat, dieses Feld mit dem Horizont, das sich immer weiter erstreckt, wie Husserl sagt, und in immer mehr unbestimmten Gegebenheiten oder Vermeintlichkeiten dem Wahrnehmenden doch selbst gegenwärtig ist. Ich sehe mir gegenüber einen Herrn sitzen; er ist aber von anderen sitzenden Damen und Herren, auf die ich schon nicht mehr achte, umgeben, bis an den Rand des Saales, bis an die Wände, die mir die weitere Welt jetzt verdecken. Durch die Wände kann ich nicht hindurchblicken. Aber dieser Saal ist doch nicht alles, was für mich gegenwärtig ist. Dahinter, also hinter dieser Mauer, ist noch freie Luft, der freie Platz der Universität, dann gibt es diese Brunnen, und das Monument; um den Platz herum stehen mehrere Universitätsgebäude, das Administrationsgebäude usw. und dahinter gibt es irgendwelche Felder und dann noch die Stadt, es geht weiter bis an das Meer. Wenn man es will geht es über das Meer, immer mehr unbestimmt, mehr oder weniger mitgegeben oder nicht mitgegeben, nur angenommen. In dem Sinne also wird das, was gerade wahrgenommen wird und natürlich aufmerksam erfasst wird, aus diesem ganzen Wahrnehmungsfeld, diesem ganzen Horizont herausgehoben. Und dieser Horizont ist unendlich. Das ist gerade unsere Mitwelt oder Umwelt oder Welt, die zu jeder Wahrnehmung gehört. Bei Husserl gehört es zum Wesen der äusseren Wahrnehmung, dass immer dieses Feld und diese Umgebung und dieser Horizont ins Unendliche weiter, bis an die anderen Galaktiken, weiter gehen. Es ist eine gewisse Rekonstruktion dessen, was tatsächlich bei Husserl vorkommt. Es scheint wichtig diese zu betonen, weil – wie bekannt – 17 Jahre nach den *Ideen*, das Werk *Sein und Zeit* von Heidegger erschienen ist, wo der Begriff des "In-der-Welt-sein" geprägt wurde. Dieses "In-der-Welt-sein" wird bei Heidegger und viel mehr bei Schülern von Heidegger als eine besondere Entdeckung behandelt, die gewissermassen gegen Husserl gerichtet ist und die einen Weg zu einer realistischen Entscheidung bilden soll. Dieses "In-der-Welt-sein" ist bei Husserl, abgesehen von dem neuen Wort, eine *alte* Geschichte; es bedeutet dass jede Wahrnehmung eine Hervorhebung aus einem Wahrnehmungsfeld, mit einem Horizont und einer Umwelt ist. Es kann nicht anders sein, und der Wahrnehmende ist in dieser ganzen Sphäre von vornherein als Wahrnehmender enthalten. Vom Husserlschen Standpunkt ist das "Inderweltsein" gar keine Neuheit; es ist die Weise des Seins des wahrnehmenden Menschen. Zumal damit die idealistische Vorbestimmung seines Gedankenganges zusammenhängt, die darüber entscheidet dass es bei Husserl gar keinen Ausweg, gar kein Mittel, aus dem Pfad zum Idealismus herauszukommen gibt. Er hat seine Argumente. Husserl selbst behauptet dass der Weltbegriff wiederum nur eine Phase der Konstitution, das heisst, nur ein höherer Typus der intentionalen Gegenständlichkeit ist; das ist die Welt, in der man als Wahrnehmender immer da ist. Trotzdem, der Einzelne als reines Ich ist derjenige, der das alles in die Mannigfaltigkeit von Erfahrungen versetzt. Das In-der-Welt-sein, das wäre nach Husserl einfach eine andere Formulierung gerade dieser Tatsache, dass die äussere Wahrnehmung, jede Wahrnehmung, ein Hervorheben aus einer mitgegebenen Welt ist.

Jetzt beginnt aber diese schwierige Diskussion, wobei manches weiter entwickelt sein muss.

Es klingt zunächst fast wie ein Paradoxon bei Husserl, als ein Widerspruch gegen die ursprüngliche Thesis. Denn der Ausgangspunkt für die Analyse der äusseren Wahrnehmung bildet die Feststellung, dass in der Wahrnehmung das Ding *selbst*gegeben wird, in seiner eigenen Person. Selbst, das heisst auch: direkt gegeben. Es ist anwesend, präsent. Aber später sagt Husserl auf einmal in den *Ideen*: Jedes Ding, oder jeder Vorgang an Dingen, wird wahrnehmungsmässig gegeben in Erscheinungen, in Abschattungen, in Ansichten, oder durch Erscheinungen, durch Abschattungen, durch Ansichten. Es wird zwischen dem wahrgenommenen Ding, dem wahrgenommenen Gegenstand und den sogenannten Abschattungen unterschieden, die früher bei Husserl "Aspekte" hiessen. (In den *Ideen* heissen sie auch Ansichten, Erscheinungen.) Es scheint also, dass da wiederum ein Schritt von dem offenen Bewusstsein in das geschlossene Bewusstsein zurück gemacht wird. Das, was direkt da ist, sind die Erscheinungen, die Ansichten, die Abschattungen. Und durch diese Welt der Abschattung wird jetzt das Zweite, das heisst das wahrgenommene Ding zur Anschauung, zur Gegenheit gebracht. Es liegt dabei nahe zu sagen, dass das, was direkt da ist, die Ansichten sind; dass das direkt Gegebene, das heisst das Erfasste, die Ansichten sind (die alten "Ideen" bei Locke), und dass das, was Ding ist, das wahrgenommene Ding, nur durch die Erscheinungen, durch die Ansichten, und durch die Abschattungen dargestellt wird. Kurz, es stimmt nicht, dass in der Wahrnehmung etwas *selbst gegeben* wird, dass insbesondere die Dinge *selbst gegeben* werden. Husserl würde gegen diese Auslegung protestieren und daran festhalten, dass die Dinge leibhaftig gegeben, selbstgegeben, sozusagen *in persona* sind. Und sie sagen, dass die Ansichten, die Abschattungen überhaupt nicht "gegeben" werden, dass sie nicht wie Dinge zur Gegebenheit "gebracht" werden. In demselben Sinne wissen sie nichts von den Abschattungen, von den Ansichten, wenn sie das Ding sehen.

Einige jüngere Freunde von Husserl haben behauptet diese Ansichten, diese Anschattungen, diese Erscheinungen seien wiederum eine psychologische Erfindung: so etwas gibt es nicht! Husserl würde darauf antworten dass, wenn wir Dinge sehen, *es so etwas* wie die Abschattungen oder Empfindungen für uns nicht gibt. Das bedeutet aber nicht, dass es sie überhaupt nicht gibt.

Wir werden noch einmal versuchen an einigen Beispielen den Unterschied zwischen dem wahrgenommenen Ding und der Ansicht, bzw. der Abschattung zu verdeutlichen, daraus zwei verschiedene Situationen machen. Wenn ich das jetzt tue, bin ich schon der reflektierende Philosoph, der etwas, das im normalen Falle spontan vor sich geht, wenn er es wahrnimmt, nun schon in gewissem Sinne dekonstruiert. Es ist tatsächlich so, wenn ich diese Bank sehe so wie sie aus Holz und schon gelblich ist und an diesem Platz im Raum steht, dass mir dann die Ansicht von dieser Bank nicht gegeben wird, ich nehme sie gar nicht wahr. Was ist mir also gegeben, wenn ich die Bank sehe? Beschränken wir uns auf ganz einfachen Details. Der Sitz dieser Bank ist seiner Form nach ein Rechteck und zwar ein Rechteck, das etwas schräg im Raume aufgestellt ist. Der Sitz hat natürlich ein Volumen, er ist etwas dick, aber seine Oberfläche ist rechteckig. Wenn man aber das Rechteck in einer Zeichnung wiedergeben will, dann muss man aufs Papier ein Trapez oder ein Rhomboid oder einen Rhombus malen, damit man im Bild die rechteckige Form z.B. einer Bank im Raum in einer bestimmten Lage sehen kann. Das sind eben Tatsachen, die die Theorie der Perspektive erfasst hat, und zwar der Perspektive, die von der europäischen Kunst im 15. und 16. Jahrhundert entwickelt worden ist. Es gibt auch eine fertige mathematische Theorie darüber, es gibt Gesetze, die genau diese Logik erklären. Wenn man die rechteckige Bank von oben, so im Raum, in dieser Lage, *sehen* will, dann muss man zugleich ohne dass man darauf achtet, nicht eine rechteckige Gestalt, sondern gerade dieses Trapez oder diesen Rhombus oder Rhomboid usw. erleben. Nach Husserl wird das von mir erlebt; es ist irgendwie da, aber es bleibt, währenddem ich das Ding sehe, für mich bis zu einem hohen Grade verborgen. Wenn ich Analytiker bin, dann enthülle ich das erlebte Trapez, ich entdecke es sozusagen. Aber dann ist mir im Grunde diese rechteckige Bank in der Wahrnehmung nicht schlicht gegeben. Ich kann jetzt diese ganze Sachlage gewissermassen umkehren. Ich gehe also von der Wahrnehmung der Bank auf die erlebte Ansicht von ihr zurück, und dann wiederum von der zum Bewusstsein gebrachten Ansicht zu dem Wahrnehmen der Bank zurück. Auf diese Weise sehe ich die rechteckige Bank wieder. Man kann das aber auch anders erfassen, und dann sieht man nicht die rechteckige, sondern die schräge Linie vor sich. Die Form der Ansicht ist jetzt eine nicht rechteckige, jedoch das wahrgenommene Ding *hat* eine rechteckige Gestalt.

Ein anderes Beispiel: Wir haben eine schöne, glatte, rote Kugel vor uns. Beim Wahrnehmen muss man die rote Kugel sehen und zwar als *rote* Kugel. Wenn man die Kugel aber zeichnen will, würde man vielleicht, wie es die Kinder tun, versuchen einen Kreis zu malen, dessen innere Fläche mit demselben Farbstift einheitlich rot bemalt würde. Das Ergebnis auf dem Papier aber wäre jetzt nicht eine Kugel, sondern lediglich so etwas wie ein Halt-Signal bei der Eisenbahn: eine rote Fläche mit weissem Ring. Um die Situation zu klären, kehrt man wiederum zu der erlebten Ansicht zurück, die man sich zunächst nicht klar zum Bewusstsein gebracht hat. Dann findet man natürlich, dass der angebliche Kreis im Innern vielfarbig ist, und zwar ist er immer noch rot, aber da er auf verschiedene Weisen "beschattet" wird, ist er von dieser Seite dunkelrot und von jener Seite, auf der er nicht beschattet, sondern gerade beleuchtet wird, da ist er hellrot. Das Rot dominiert, durch den Schatten und durch die Beleuchtung; aber es gibt da noch etwas: diese glatte Kugel glänzt an einer

Stelle. Weiterhin sagen wir dann, eine rote *Kugel* hat eine ganz bestimmte dreidimensionale Gestalt. Liegt sie z.B. auf einer Fläche, wo es gelbe oder grüne Gegenstände gibt, dann erscheint jetzt wiederum ein neues Element; es gibt "Farbenreflexe", die "grünlich" oder "violett" usw. sind. Will man das jetzt im Bild rekonstruieren, dann muss man viele Farben benutzen und die Farbenflecke entsprechend auf dem Papier verteilen. Dann sieht man aber nicht nur einen roten, flachen Kreis, sondern eine Kugel. Den Kreis kann man zwar auf dem Papier auch wahrnehmen, aber er ist dann vielfarbig.

Man kann sich zu der Zeichnung auf dem Papier auf doppelte Weise einstellen: entweder sieht man einen Kreis, einen flachen Kreis, der mit verschiedenen Farben künstlich bedeckt wird, oder man sieht eine kugelige Gestalt, die einfarbig ist. Wir können wiederum der Ansicht von einer roten Kugel die wahrnehmungsmässig gegebene Kugel, die einfarbig und kugelig ist, gegenüberstellen. Die eintönig "rote" Kugel wird aber als von einer Seite beleuchtet und glatt gesehen, während sie von der anderen Seite beschattet ist. Das ist in der Wahrnehmung auch gegeben. Noch etwas ist jetzt aber zu beachten! Wenn diese Kugel auf einem Billardtisch von mir wegrollt, dann sehe ich, wenn ich sie wahrnehme, trotzdem immerfort dieselbe gleich grosse Kugel. Wenn ich sie aber mit einem Filmapparat aufnehmen würde, dann würde es sich zeigen, dass in den späteren Aufnahmen die Kugel immer kleiner erscheint. Wenn ich bei der Wahrnehmung der sich entfernenden Kugel zu erlebten Ansichten übergehe, dann wird in ihnen die Grösse der Kugel immerfort verändert; sie wird immer kleiner und ist ganz klein geworden, wenn sie ein paar Meter von mir entfernt liegt. Die Photographie rekonstruiert jedoch mit Hilfe von technischen Mitteln die Momentbilder.

Die Malerei und zwar der Impressionismus hat uns gelehrt, wie mannigfaltig hinsichtlich Farbe, Intensität, Licht, Beweglichkeit usw. all das ist, was wir bei einer visuellen Wahrnehmung eines Dinges als seine wandelnden Ansichten erleben. Das Merkwürdige ist, dass man das Bild, das impressionistische Bild, doppelt interpretieren kann. Man kann es sehen als eine Flache Oberfläche mit verschiedenen "Flecken", und ohne Gegenstand, oder man kann eben diese "Flecke" sozusagen als Ausgangspunkt für eine besondere Wahrnehmung machen und dann durch diese Mannigfaltigkeit von verschiedenen Farben Dinge "sehen". Die alten Maler haben Dinge in ihren objektiven Eigenschaften gemalt, wobei sie natürlich gezwungen waren, gewisse Ansichten im Bilde zu rekonstruieren. Die Impressionisten indessen rekonstruieren im Bilde in erster Linie nicht Dinge, sondern Ansichten. Dies sind nicht Ansichten, die schon einen gegenständlichen Sinn haben, sondern Farbendaten, die den von uns erlebten "Empfindungsdaten" entsprechen. Sie rekonstruieren also nicht etwas, wie diese Gestalt, die nicht rechteckig, sondern trapezoid oder rhomboid ist. Das, was sie mit Hilfe von technischen Mitteln rekonstruieren, was sie auf dem physischen Ding, das Leinwand heisst, mit Farbenpigmenten hervorbringen, wird gemacht, damit wir als Betrachter etwas haben, was hinter oder unter der Ansicht noch versteckt ist und was wir uns noch nicht zum Bewusstsein gebracht haben. Das ist, was man jetzt nach Husserl "Empfindungsdaten" nennt.

Diese Empfindungsdaten sind nicht die Flecken, die objektiv auf dem Papier sind, nicht die Farbenflecke, es ist vielmehr die sozusagen untere, lebendig erlebte, an sich *quasi* versteckte, fliessende Mannigfaltigkeit von farbigen Daten, die nach den Impressionisten, meines Wissens als erster Henri Bergson beschrieben und sie die "données immédiates de la conscience" genannt hat. Später hat er sie in dem Buch *Matière et mémoire* als "continuité hétérogène" beschrieben. Er spricht auch von "sensation". In dieser Terminologie "continuité", aber zugleich "continuité hétérogène", gibt es gar keine Abgrenzungen zwischen den einzelnen Daten, einzelnen Farben, Abschattungen, wie Husserl sagen würde. Alles ist irgendwie verschwommen, fliesst zusammen; es gibt höchstens das, was Husserl "Sprung in der Qualität" nennt, aber es gibt da keine Konturen, keine Vielheiten. Das Feld der jetzt von mir erlebten Empfindungsdaten ist ein Ganzes, von dem ich, sobald ich den Blick daran abwende, gar nichts mehr weiss. Ich habe darauf nicht geachtet; wenn ich das fassen will, dann verschwindet mir der ganze Saal, die Menschen verschwinden mir, auch die Ansichten mit ihrem besonderen Sinne. In jedem Akt, in jeder Ansicht gibt es so eine letzte qualitative Gestalt; das sind die so genannten "Empfindungsdaten", die sich stets wandeln.

Auf welche Weise entdecke ich sie eigentlich, wenn ich keine besondere Analyse mache? Auf die Weise, dass, wenn ich z.B. ganz ruhig in den Saal blicke und die Anwesenden sehe und das eine Weile dauert, dann ist es auffallend: was ich jetzt erlebe, ist niemals konstant; es ist in jedem neuen Augenblick schon etwas anders; es ändert sich in Bezug auf die Intensität der gerade erlebten Daten; es vollzieht sich immer irgendeine

Wandlung im Feld. Ich kann die ganze Zeit meine Augen und meinen Kopf unbeweglich halten, doch was immer ich erlebe ist im Fluss und es spricht mich mehr oder weniger bei der Wahrnehmung an. Manchmal verschwindet es ganz, und manchmal ergibt sich etwas Neues. Wenn ich z.B. ein Automobil lenke, dann brennt rechts eine Lampe und bewegt sich, und ich sehe hin. Aber das heisst: Zunächstmal hat sich im Feld meiner Empfindungsdaten etwas geändert, was mich angezogen hat; dann muss ich das jetzt wahrnehmen, und dann stelle ich fest, dass es die Lampe ist, die da brennt. Sobald ich Dinge sehe, dann sind die Dinge für mich abgegrenzte Ganzheiten. Dies wird freilich nicht besonders von mir unterschieden, der Herr soundso und Frau soundso, die Bank soundso, Papier soundso erscheinen mir abgegrenzt. Wenn ich schon zu den Ansichten zurückgehe, dann ist das auch noch irgendwie abgegrenzt; also es ist diese Gestalt, die nicht rechteckig, sondern gerade rhomboid oder gar rund ist. Wenn ich das im Bild zeigen will, dann kann man das auf verschiedene Weise tun. Ich zeichne Konturen oder ich bringe Kontraste hervor, so dass sich das Eine von dem Anderen abgrenzt und sich die neue Gestalt, die erlebt wird, für mich konstituiert. Wenn ich jetzt versuche, in das Feld meiner ursprünglichen Daten – Empfindungsdaten – zurückzudringen, sie irgendwie zu fassen, dann zeigt es sich, dass alles eins ist, dass das Ganze eins ist, es ist eben "continu", es gibt gar keine Abgrenzungen, es besteht nichts Konstantes, es fliesst, immer neu, bewegt sich.

Jetzt entstehen jene grossen Fragen, die in der Neuzeit, im Grunde zuerst Kant aufgeworfen hat. Denn Kant hatte auch diese Empfindungsmannigfaltigkeiten entdeckt. Der menschliche Geist hat eine besondere Form, in der die Empfindungsmannigfaltigkeiten erfasst werden. Raum und Zeit sind die beiden Anschauungsformen, in welchen die Empfindungsdaten ausgedrückt werden. Dazu kommt das System der Kategorien, die eine weitere Formung des gelieferten Materials durchführen. Auf diese Weise kommt man von den einfachen, ursprünglichen Daten zu den Erscheinungen im Kantischen Sinne. Die "Erscheinungen" sind bei Kant einfach die anschaulich gegebenen Dinge, Kant sollte im Sinne seiner Theorie sagen: in dem ursprünglichen Feld der Empfindungen (im Kantschen Sinne) gebe es gar keine räumliche Struktur, gar keine zeitliche Struktur und gar keine kategoriale Formung. Und wenn Kant die tranzendentale Deduktion der Kategorien durchführt, sollte er in dieses

Gebiet zurückkehren, wo reine Empfindungsdaten sind, wo es noch keine Kategorien gibt, keinen Raum, keine Zeit, weil das alles bei Kant subjektiv *a priori* ist. Erst von da aus sollten die "Kategorien" "deduziert" werden. Aber Kant kann das nicht machen, er kann in dieses formlose Gebiet der Empfindungen nicht vordringen, weil es ihm seine eigene Theorie der apriori'schen Formen verbietet. Er kann keinen Zugang zu den ursprünglichen Daten haben, weil der Notwendigkeitscharakter der apriori'schen Formen das unmöglich macht. Das Empfindungsmaterial muss unmittelbar in Zeit- und Raumformen und in Kategorien s.z.s. erfasst werden. Dies ist die einzige Möglichkeit, dieses Material "rein" zu haben. Worin liegt der Schritt zur Befreiung aus dem Käfig, den Bergson unternimmt? Für Bergson sind die Kategorien keine "apriori'schen" Formen; es gibt für ihn auch keine notwendigen "apriori'schen" Raum-, Zeit-, Anschauungsformen. Dennoch kann ich zu den ursprünglichen Daten, zu der fliessenden "continuité hétérogène" selbst vordringen; ich kann die "données immédiates de la conscience" erreichen. Der ganze eventuell noch vorhandene Rest sind gar keine *apriori'schen* Formen. Es sind lediglich Strukturen des Intellekts, und der Intellekt ist auf die Handlung bezogen relativ. Intellekt ist, nach Bergson, diese Macht in mir, die die Welt, mit der ich ursprünglich verkehre, – das heisst die Welt der "données immédiates", durch Formung verwandelt und verfälscht. Die "wirkliche" Wirklichkeit, ist der Strom der einen Daten, der ursprünglichen Daten in der reinen Dauer. Dazu kommt noch die Theorie des Intellekts, die Bergson in *Matière et mémoire* durchgeführt hat, und zwar ohne jede Metaphysik wie das später in der *L'évolution créatrice* der Fall ist.

Unter anderem Bergsons sehr geistreichen Analysen des Überganges von der "perception pure" zu der "perception concrète" sind bemerkenswert. "Perception pure" ist diejenige Situation, in welcher ich nur die fliessenden "sensations" habe, und an gar keiner Handlung interessiert bin. Ich höre auf, ein Handlungszentrum zu sein, ich kehre um zu der ursprünglichen Wirklichkeit (im Sinne Bergsons) und dann kann ich die "données immédiates" verfolgen, wie ich von ihnen zu der konkreten Perzeption und zu den wahrgenommenen Dingen komme. Dinge sind dennoch nichts anderes als gewisse Gebilde, die durch die Schemata der Handlung geformt sind. Diese Schemata, insbesondere ihr homogener Raum und die Kategorien sind intellektuelle Formen, die auf die Handlung bezogen relativ sind. Sie werden auf die ursprüngliche fliessende

Wirklichkeit angewandt und entstellen sie. Da sie aber nicht "a priori", sondern auf die Handlung bezogen relativ sind, können sie prinzipiell beseitigt werden. Wenn man reine, uninteressierte Erkenntnis haben will, muss man diese Schemata von sich abschütteln, dann kehrt man zu der reinen Dauer, zu der ursprünglichen "continuité hétérogène" der ursprünglichen Daten.

Die Bergson'sche Theorie des Intellekts und der Handlungsrelativität der "Kategorien" ist nicht haltbar, wie ich in meiner Abhandlung "Intuition und Intellekt bei H. Bergson" zu beweisen suchte. Die Entdeckung der "donnés immédiates" besteht aber zurecht, so wie eine Reihe von Beschreibungen, welche Bergson über sie gibt. Da erhebt sich das wichtige Problem: Sind die ursprünglichen Empfindungsdaten zunächst im Gebiet der visuellen Wahrnehmung, in jedem Sinne vollkommen begrenzt, wie sie sein müssten, wenn Kant recht hätte, dass Raum eine apriorische Anschauungsform ist. Auch diejenigen, welche die sog. "Empfindungen" für etwas "bewusstseinsmässiges" und zugleich jegliches Bewusstsein für schlechthin begrenzt halten, müssen für die Begrenzung der ursprünglichen "unmittelbaren" Daten eine Erklärung abgeben. Demgegenüber sagt Bergson – und so heisst es im Sinne Husserls – nur dass bei ihm das Problem etwas anders beleuchtet wird: es gibt "une étendue concrète de la sensation", es gibt eine konkrete Ausdehnung; Sichdehnen, der ursprünglichen, erlebten Daten. Raum, "c'est l'espace", der homogene Raum, ist eine intellektuelle, mathematische Konstruktion, ein abstraktes Gebilde, das sich aus den Grundtendenzen der intellektuellen Erkenntnis ergibt. Er ist in den ursprünglichen "données immédiates" einfach nicht vorhanden. Aber das wirklich Erlebte, diese Daten, sind "étendues", sind ausgedehnt, haben eine ursprüngliche Ausdehnung. Husserl sagt ungefähr dasselbe, obwohl er Bergson gar nicht kannte; er hat es selbst beschrieben.

Es ist interessant festzustellen, dass von zwei verschiedenen Philosophen dasselbe entdeckt wurde.

Und wie ist es bei den ursprünglichen Empfindungsdaten mit der Zeit? Bergson würde sagen (und Husserl auch), diese Daten liegen nicht in der Zeit und zwar in der konstituierten homogenen Zeit. – Nach Bergson ist übrigens die homogene Zeit dasselbe, wie der homogene Raum, nur dass sie ein eindimensionales *continuum* ist, während der Raum (l'espace) ein mehrdimensionales mathematisch bestimmtes *continuum* ist. – In den ursprünglichen Erlebnissen gibt es diese mathematisch bestimmte homogene Zeit nicht. Die ursprünglichen Daten indessen "dauern"; im Sinne von Bergson ist es die ursprüngliche Dauer, la "durée pure"; sie ziehen sich in gewissem Sinne in die Länge. Wenn man sagt "ziehen" sich in die Länge, so ist das vom Standpunkt Bergsons schon eine gewisse Verfälschung: wo von einem "Ziehen" die Rede ist, wird irgendwie ein Raummoment eingeführt, während "la durée pure" gar nicht ausgedehnt ist. Auch bei Husserl ist das ursprüngliche, innere Zeitbewusstsein, wie er sagt, nicht "étendu", es ist nicht "ausgedehnt", hier gilt "Dimension". Wo aber Bergson ausserhalb der reinen Dauer nur die "Zeit" (le temps) im Sinne eines eindimensionalen homogenen *continuum* sieht und sie mit dem "Raum" (*l'espace*) identifiziert, unterscheidet Husserl neben dem ursprünglichen inneren Zeitbewusstsein einerseits die "konstituierte" erfüllte, qualitativ bestimmte Zeit, andererseits die "physikalische" mathematisch bestimmte kontinuierliche Zeit, die Zeit der physikalischen Geschehnisse. Die "erfüllte" qualitativ bestimmte Zeit ist die Zeit der konstituierten Erlebnisse und auch der konkreten, in der äusseren Wahrnehmung gegebenen physischen Geschehnisse in der realen Welt. Nur die ursprünglichen Empfindungsdaten, die aller Konstitution der inneren und äusseren Geschehnisse zugrundeliegen, sind von der konstituierten "erfüllten" Zeit frei. Sie "dauern" aber – gerade so wie es Bergson behauptete. Es gibt ein ursprüngliches, primitives "étendue", ein Sichausdehnen. Es hat genauso Sinn darüber z.B. bei ursprünglichen, visuellen Daten zu sprechen, so wie es Sinn hat auch bei den akustischen Daten über diese gewisse "Dehnung" zu sprechen. Auch die taktuellen Daten sind nicht ganz ohne jede Spur der Ausgedehntheit. Es ist auch sinnvoll zu sagen, dass diese Daten eben "dauern" und im Dauern sich auf eine ganz bestimmte Weise verwandeln, auftauchen, aktuell werden und abklingen. Die Frage aber kommt auf: Wie ist es bei den ursprünglichen Empfindungsdaten mit den Kategorien? Nicht alle zwölf Kategorien Kants sind wirklich formale Strukturen. Es werden verschiedene angebliche "Kategorien" genannt, die gar keine Kategorien, also ursprünglich gegenständliche Formen, sind. So ist z.B. die Realität gar keine "Kategorie". Es ist eine bestimmte Seinsweise, aber keine formale Struktur, die in der existentialen Ontologie zu klären ist. Auch "Notwendigkeit" ist keine "Kategorie", keine ursprüngliche gegenständliche Form.

Gibt es in dem Gebiet der ursprünglichen Empfindungsdaten überhaupt keine "Form", also gar kein "Subjekt" von etwas-sein; gar keine "Bestimmtheit von etwas-sein", gar keine "Identität"? Bergson würde das behaupten, da er das Vorhandensein jeglicher "Form" in der Sphäre der reinen Dauer leugnet. Ich habe darüber in meinem Buch über Bergson eine Diskussion durchgeführt und behauptete, dass Bergson doch im Widerstreit mit sich selbst ist und dass er dazu noch eine *petitio principii* begeht. Es ist, so behaupte ich, nicht möglich, die Kategorien auch in diesem Feld der ursprünglichen Daten zu leugnen. Tut man es, dann begeht man einfach ein Irrtum, Husserl hat sich nicht darüber geäussert, wie es mit den Kategorien in den Empfindungsdaten steht. Er hat aber gegen die These meiner Dissertation nicht protestiert. Aber da ist er vielleicht nur der Lehrer, der dem Schüler Freiheit lässt. Die kategoriale Erfassung der ursprünglichen Empfindungsdaten soll hier also als ein Problem bestehen bleiben.

Es gibt aber noch ein anderes, prinzipiell wichtiges Problem. Es fragt sich nämlich, in welcher Beziehung diese ursprünglich erlebten Empfindungsdaten, die hinter den fliessenden Ansichten oder auf deren Grund doch irgendwie für mich da sind, zu meinen Bewusstseinsakten und zu mir selbst stehen. Bei Kant war die Sache folgendermassen: die Empfindungen sind Effekte der Affizierung, welche die Dinge an sich auf uns ausüben. Sie sind subjektiv und befinden sich irgendwie im Bewusstsein, im Kant'schen Bewusstsein. Wie es damit für Bergson stand, ist sehr schwer zu sagen. Natürlich, "la continuité hétérogène" des "sensations", das ist die letzte ursprüngliche Wirklichkeit, und zwar, wie es scheint, "de la conscience". Bergson hat aber im Grunde gar keine Intentionalität des Bewusstseins, gar keine dieser Art "Akte", von denen wir im folgenden sprechen werden. Man kann nicht sagen, ob diese "continuité hétérogène", die *données immédiates*, also die ganze Mannigfaltigkeit des in meinem Leben "Empfundenen", dasjenige etwas, was ich selbst bin, ob ich es bin, oder ob es etwas ist, was ich einfach nur vorfinde, wozu ich irgendwie gekommen bin, das ich eben "empfinde". Bin ich selbst aber nicht die Mannigfaltigkeit der "données immédiates"? Um diese Frage etwas konkreter zu fassen, muss man zunächst ein paar Worte über die Akte sagen. (Siehe die Tafel).

Die ursprüngliche Struktur des Wahrnehmens lässt sich etwa so darstellen: Wir beginnen mit dem Ich und dieses Ich lebt und vollzieht eine Wahrnehmung, das heisst es wird etwas von mir wahrgenommen und

Ich	Akt-Intention Erleben der Ansichten "Durchleben" des Aktes im Akt selbst enthalten.	Ding (Gegenstand) Ansicht (Abschattung) "Empfindungsdaten" besondere Struktur des Aktes "Selbstbewusstsein" (Kant); Brentano: "inneres Bewusstsein"	Das Ding "gegeben" Ansichten "erlebt" Akte "durchlebt" vom Ich

erfasst: es ist mein intentionaler Akt, der sich auf etwas bezieht. Es bezieht sich auf das gegebene Ding. Also was erfasst wurde, ist das gegebene Ding in seinen eigenen Bestimmtheiten. Wenn ich wahrnehme, so ist es für mich bloss ein dies und das. Ich vollziehe den Akt, aber ich achte nicht auf meine Wahrnehmung, ich vollziehe sie nur. Das Ding ist gegeben. Damit ich es aber "sehen" kann, muss ich eine Abschattung, dh. eine Ansicht erleben, die mir nicht gegeben ist, die aber wenn *das* Ding gegeben ist, für mich nicht verschwindet. Ich muss doch irgendeinen Zugang auch zu der Ansicht haben. Wir unterscheiden zwischen dem was "gegeben" wird und dem was "erlebt" wird. Dieses Erleben, das ist sozusagen eine zweite Form, eine zweite Weise des Bewusstseins beim Wahrnehmen; es ist kein Treffpunkt einer Intention. Der Treffpunkt der Intention – das ist das Ding, oder eine Eigenschaft des Dinges usw. aber die "Ansicht" (Abschattung) wird mir doch irgendwie gegenwärtig, anwesend, ich werde davon irgendwie berührt. Es ist jene früher erwähnte Gestalt des Rhomboids, wenn ich ein Rechteck im Raume sehe. Dahinter gibt es jetzt jene fliessende Mannigfaltigkeit von Daten, die ich nicht in dem Sinne "erlebe", wie die Ansichten, sondern "empfinde". Dabei werde ich durch diese Daten auf ganz bestimmte Weise, wie Husserl sagt, von ihnen irgendwie "affiziert", "berührt". Ich werde "affiziert" und infolgedessen wende ich mich ihnen zu, aber dann vollziehe ich gleich einen Wahrnehmungsakt. Und da hat sich mir sofort irgendeine Ansicht konstituiert, ich sehe das Ding und erfass es in seinen Eigenschaften. Das Ding ist es was erfasst wird und nicht die Empfindungsdaten. Wenn ich eine Gegebenheit der Empfindungsdaten erzielen will, dann muss ich eine Reflektion vollziehen oder eine Analyse durchführen, so wie ich das jetzt tue. Dann aber deformiere ich bis zu einem gewissen Grade den ursprünglichen Gang der fliessenden empfundenen Daten.

Das Empfinden der ursprünglichen Daten ist nicht die Intention, in der nun kein intentionaler Gegenstand gegeben wird. Der intentionale Gegenstand ist das Gegebene, und was da bloss "empfunden" wird, ist für mich irgendwie ein Neues, gerade weil es auf eine ganz bestimmte Weise empfunden und nicht wahrgenommen wird. Die Frage stellt sich, ob der Akt des Wahrnehmens, der Akt des Aufmerksamseins, des Sichaufetwasrichten nicht wiederum eine Konstruktion ist, die nicht durch Intuition belegbar ist. Darauf lässt sich antworten dass dieser Akt und auch dieses Empfinden wiederum etwas ist, wovon ich wissen kann, ohne zu reflektieren, ohne reflektieren zu müssen. Ich kann natürlich darauf reflektieren, ich kann nämlich auf diesen Akt einen neuerlichen zweiten Akt richten, das heisst Reflektion, Introspektion vollziehen. Diese "Reflektion" ist die "immanente Wahrnehmung", von welcher Husserl spricht; und wir werden sofort darauf eingehen. Aber dann ist das Ganze, der Akt, welcher ein so bestimmtes Ding trifft und zugleich Ansichten erlebt und ursprüngliche Daten empfindet – dieses Ganze ist jetzt Korrelat eines neuen Aktes, den man "innere", die "immanente" Wahrnehmung nennt. Es gibt eine ganz bestimmte (Husserl sagt es auch) Spaltung zwischen dem vollzogenen, das Ding erfassenden Akt und dem reflektierenden Akt der immanenten Wahrnehmung.

Jetzt aber entsteht der Streit zwischen dem Neukantismus und Husserl. Die Frage nach dem Ich drängt sich auf: Wenn ich eine Reflektion vollziehe, dann verschwindet gewissermassen das Ich, dass den Akt vollzieht, auf welchen die Reflektion gerichtet ist; ich befinde mich schon irgendwie auf einem anderen Niveau und ein anderes Ich wendet sich reflektierend auf dieses Ich, obwohl die Tendenz des Ich dahin geht, in der Reflektion sich selbst zu erfassen. In dieser Sachlage behaupten die Neokantianer dass das Ich, das reine Ich überhaupt nicht erfassbar wäre. Bei allen Versuchen es reflektiv zu erfassen, entschwindet es. In der Reflektion wäre es nicht das (reflektierende) Ich, sondern nur ein besonderer Gegenstand: das gewesene Ich, das z.B. die äussere Wahrnehmung vollzogen hat, auf welche sich die Reflektion richtet. Das echte reine Ich, dasjenige, das die Reflektion vollzieht, muss nur vorausgesetzt werden, es ist in seiner Ich-haften Position und Funktion nie zu erfassen.

Husserl vertritt eine entgegen-gesetzte Auffassung seit der II. Auflage der *Logischen Untersuchungen* (und später). Er behauptet, dass es möglich sei, das reine Ich zu erfassen, ohne es seiner eigenen Funktion und Stellung zu berauben. Wie kann man das aber feststellen? Man kann, wie einst auch Brentano vorgeschlagen hat, das sogenannte "innere Bewusstsein" annehmen, welches in dieser Sachlage behilflich sein kann. Es ist aber nicht ganz klar, was Brentano darunter verstand. So lange aber die Stellung des inneren Bewusstseins nicht geklärt wird, gilt der Zweifel, woher ich weiss, dass ich jetzt so ein Doppelleben führe, dass ich einerseits diesen Akt vollziehe und das Ding wahrnehme, und zugleich darauf reflektiere. Ich kann darüber hinaus eine neue immanente Wahrnehmung einführen, welche s.z.s. die Gesamtsituation erfasst. Wenn ich dann noch Zweifel habe, ob ich diese meine Wahrnehmung wirklich vollziehe, dann kann ich eben meine immanente Wahrnehmung zu Hilfe nehmen, die sich auf die Wahrnehmung des Wahrnehmens usw. richtet. So erhalten wir aber einen Regressus ins Unendliche. An dieser Stelle kann ich wirklich nicht mehr sagen, ob ich das alles wirklich auf Grund einer Wahrnehmung weiss, oder ob ich das nur so konstruiere.

Ich gebe zu, dass es so etwas wie immanente Wahrnehmung gibt. Wird aber die immanente Wahrnehmung gebraucht um wirklich zu wissen, dass ich etwas erlebe? Ehe ich diese "innere", diese immanente Wahrnehmung zu vollziehen beginne, ist sie mir schon bewusst, meine immanente Wahrnehmung verspätet sich in gewissem Sinne, sie erfasst etwas, was schon vorbei ist. In gewissem Sinne sie ist bloss wie ein Nachklang. Woher weiss ich, dass ich vor dem Moment, wo ich zu reflektieren beginne, schon bewusst gelebt habe und mich nicht auf mich selbst zurückbesinne? Ich brauche nicht auf mich zu reflektieren, ich brauche gar keine immanente Wahrnehmung zu vollziehen, ich kann darauf verzichten, denn dieser Akt, der erste des Wahrnehmens, das ist schon eine ganz besondere Weise zu leben, das ist das *Sichselbstdurchdringen.* Das schlichte, bewusste Leben, z.B. ein äusseres Wahrnehmen ist – abgesehen davon, dass es ein Erfassen eines Dinges und ein Erleben einer Ansicht ist – in sich selbst kein "Erleben" mehr, sondern ein *Durchleben.* In diesem sich selbst durchlebenden Leben gibt es keine Doppeltheit; das ist im echtesten Sinne das Bewusstsein; "bewusst sein", das heisst eben bewusst leben; es heisst nicht "reflektieren", es heisst einfach schlicht wissentlich leben.

Merkwürdigerweise gibt es in der deutschen Sprache einen Ausdruck dafür, den bereits Kant in dem hier in Frage kommenden Sinne verwendet hat, obwohl man nicht bemerkt hat, dass da irgendwas Neues vermeint wird. Und zwar spricht Kant davon sowohl in der ersten als in der zweiten aber besonders in der zweiten Auflage der *Kritik der reinen Vernunft*. Er verwendet das Wort "Selbstbewusstsein". Es entspricht dem was ich hier "Durchleben" nenne. Kant sagt in der II. Auflage der *Transzendenten Deduktion der Verstandesbegriffe*: "Das 'Ich denke' muss jede meiner Vorstellungen begleiten können". Es entspricht auch dem Sinne des kartesianischen "cogito". Das "Ich denke", "cogito", bedeutet nicht, dass ich reflektiere, dass ich über mich eine Aussage fälle. "Cogito" heisst einfach und ausschliesslich "Selbstbewusstsein haben". Dieses "Selbstbewusstsein", das bei Kant auftaucht und mit Beziehung auf Descartes durch das "ich denke" ersetzt wird, ist ohne Zweifel diese ursprüngliche Seinsweise des Bewusstlebens, die darin besteht "sich selbst mit einem Wissen durchzudringen".

Es gibt gar keinen Akt, der nicht so selbstbewusst wäre; würde ihm das fehlen, so würde er überhaupt nicht bestehen. Es gibt heller und weniger hell durchlebte Akte, hellere und weniger helle Selbstbewusstseinsakte, es gibt aber gar keine völlig "dunklen" Akte. Ein derartiger Akt ist überhaupt nicht möglich, dann gäbe es überhaupt nichts.

Bei Bergson gibt es das schöne Wort "l'intuition". ("Intuition", das ist auf deutsch "Ein-sicht" oder "Ein-blick".) Leider ist dieser Ausdruck sehr vieldeutig. Es gibt einen ausgeprägten, ganz engen Sinn dessen, was eigentlich "Intuition" ist. Das ist derjenige Typus "de la conscience", der in der reinen Dauer eingetaucht ist und sich nicht "von Aussen" her in dem "statischen Aspekt" fasst. Die Intuition in diesem ganz engen Sinne, das ist nichts anderes, als jenes Durchleben der eigenen Akte.

Ausserdem gibt es bei Bergson noch einen anderen Sinn der "Intuition", der im Haben der "continuité hétérogène des données immédiates de la conscience" besteht, und dies entspricht dem Erleben der "Abschattungen" oder dem Einfinden der letzten ursprünglichen Daten. Ihm zugrunde aber liegt "la conscience", das ist eben das sichdurchlebende Bewusstsein, das "cogito" von Descartes. Descartes braucht auch nicht eine Reflektion "cogitationes" zu vollziehen. Im "cogito" begründet sich alles. Deswegen kann er auch sagen "cogito, ergo sum", denn da ist die letzte Identität des spontanen Bewusst-Lebens vorhanden.[1]

Damit kommen wir zu der Frage: was ist jetzt das Bewusstsein? Was ist der Begriff des "Bewusstseins"? Was ist die ursprüngliche Einheit des "cogito"? Die Einheit des "cogito" ist nur in einer momentanen Phase. Dann entsteht auch ein neues und schwieriges Problem: Was ist die Einheit des "cogito" (und des "ego cogito") in der Zeit und zwar in der ursprünglichen Zeit? Hier kann diese Frage nur erwähnt werden.

Nun aber wenn man die Frage aufwirft, was ist die ursprüngliche Einheit, das Ganze, was da zueinander gehört, so sind verschiedene Lösungen möglich. Zuerst kann man behaupten, dass es doch nichts anderes ist als das, was in der "Intuition" in dem engsten Sinne, im Durchleben vorhanden ist. Das ist der Akt. Dann aber bleibt ausser der Sphäre des Bewusstseins der Fluss der Empfindungsdaten, das ist das schon irgendwie Empfundene, und auch die Ansichten, das ist etwas Erlebtes, und natürlich auch die wahrgenommenen Dinge. Das alles ist schon freilich noch da, aber es ist nicht in der ursprünglichen Einheit des "cogito" enthalten. Weiterhin kann man sagen, dass die Dinge eben Dinge sind, die sich etwas ausserhalb meines Bewusstseins befinden, aber dem Bewusstsein transzendent bleiben. Was die Empfindungsdaten und die Ansichten betrifft, sagt man gewöhnlich dass es natürlich etwas Subjektives, jedenfalls nichts Dingliches sei. Das Ding mit seinen Eigenschaften bleibt ausserhalb dessen, was zum Bewusstsein gehört, es ist ihm transzendent im *engeren* Sinne, eine im Vergleich mit dem Bewusstsein zweite Einheit, ein zweites Sein. Wozu soll man aber die Empfindungsdaten hinzurechnen? Husserl behauptet sie gehören zum Bewusstsein, sie sind "reeller Teil", "reelles Bestandstück" des Bewusstseins selbst. "Reell" im Kontrast zu "real". "Reell", das heisst so viel wie "effectiv", "echt". "Reeller Teil" heisst also so viel wie effektiver Teil des Bewusstseins. Die Empfindungsdaten bilden also nur dem Akt eine Einheit, ursprüngliche Einheit.

Die Ansichten (Abschattungen) gehören natürlich auch zum Bewusstsein. Das ist schon die Konsequenz dessen, dass man die Empfindungsdaten "im" Bewusstsein hat. Denn sie sind nur das, was sich auf der Unterlage der Empfindungsdaten im Laufe der Erfahrung bzw. der fortlaufenden Wahrnehmung als ein Sinn, als eine Qualität konstituiert. Die Ansichten wachsen s.z.s. aus den Empfindungsdaten hervor. Wenn Empfindungsdaten in der ursprünglichen Einheit mit dem Akt zusammen sind, dann sind auch die Ansichten etwas, was zum Bewusstsein gehört.

Was ist nun das Ding? Die Realisten, aber auch die Menschen in natürlicher Einstellung meinen: das Ding ist ausserhalb des Bewusstseins. Husserl behauptet das ist das "cogitatum"; das ist also dem Bewusstsein transzendent. Zugleich aber sagt Husserl: das "cogitatum" als "cogitatum" gehört zu dem "cogito", es ist untrennbar mit ihm "verbunden". Es ist notwendig wie Schatten, wie eine Folge davon, dass wir solche und solche Ansichten von Etwas haben. Das ist der gegenständliche Identitätspol meiner Erfahrungen, das ist das, worauf sie bezogen sind: nur eine neue Schicht des konstituierten Sinnes. Das ist das Ding "in Anführungszeichen", wie Husserl oft in den *Ideen zu einer reinen Phänomenologie* sagt. Als "cogitatum" ist es ein Dingnoema, es ist dem Bewusstsein transzendent, es ist *kein* reeller Teil desselben, aber es ist das notwendige "pendant" dazu. Als "pendant" ist es von ihm nicht abtrennbar, wenn die Erfahrung gut verläuft. Man kann es als Schatten sozusagen nicht fortschneiden. Wenn die Erfahrung anders verläuft und unstimmig wird, wenn nicht alles zusammenstimmt im Verlaufe der fortschreitenden Erfahrung, dann "explodiert" es, – wie Husserl sagt. Dann gibt es natürlich auch kein Ding. Gibt es aber auch kein Dingnoema?

Wenn man die von mir in Husserl unterschiedenen verschiedenen Begriffe der Transzendenz zusammenstellt und fragt, in welchem Sinne da (bei dem wahrgenommenen Dinge) von "Transzendenz" gesprochen wird, drängt sich die Frage auf, ob Husserl doch Idealist sein kann, da er Transzendenzen dem reinen Bewusstsein gegenüber annimmt, – "Transzendenzen", das heisst Dinge, das heisst Kulturdinge, das heisst die Welt usw. – Ich behaupte anhand der vorigen Auslegung der Sinne der Transzendenz, dass er trotzdem Idealist bleibt. Es wird klar, dass obwohl Husserl kein Immanenzphilosoph ist, er ist trotzdem ein Idealist, weil er die transzendenten Dinge als Korrelate, als intentionale Korrelate der Erlebnismannigfaltigkeiten nicht bloss anerkannt hat, sondern sie zugleich für "untrennbar vom Bewusstsein" erklärt.[2]

II

Wie es sich zeigt, gibt es mehrere *verschiedene Begriffe von Transzendenz*, die von Husserl nicht ausdrücklich geschieden wurden. Sie müssen mit den Begriffen der Immanenz – da die Lehre vom Bewusstsein in der Sphäre der Immanenz arbeitet – klar gegenüber gestellt werden. Es kommt eben ohnedies zu verschiedenen misslichen, falschen Interpretationen. Die Begriffe, die ich da auseinanderlege oder die ich schon zu charakterisieren begonnen habe, stecken alle im Texte von Husserl, abgesehen von einem vielleicht, den ich noch am Schlusse erwähnen werde. Husserl scheidet sie aber begrifflich nicht klar genug, so dass ich zwar an jeder Stelle weiss, worum es sich eigentlich handelt, aber wenn man das nicht weiss, wenn man nicht hinter den Text zu den Phänomenen selbst geht, dann kann man sagen, dass der Begriff der Transzendenz bei Husserl vieldeutig ist, und davon ergeben sich verschiedene Schwierigkeiten.

Ich unterscheide zunächst zwei, eventuell drei *Gruppen von Transzendenzbegriffen.* Die erste Gruppe, das ist die Transzendenz im erkenntnistheoretischen Sinne. Sie ist irgendwie auf Erkenntnis bezogen. Die zweite Gruppe ist die Transzendenz im *ontischen* Sinne. Vielleicht könnte man auch sagen: im ontologischen Sinne, und da, am Rande dieser zweiten Gruppe, steht noch ein ontischer Begriff, der aber ein ausgesprochen *metaphysischer* Begriff ist, und der in der Geschichte der Philosophie und speziell in der Philosophie des Mittelalters sehr wohl bekannt ist und den man natürlich bloss diesen Transzendenzbegriffen, gegenüberstellen soll die da in Frage kommen. *Drei* von diesen Begriffen der ersten Gruppe, die bei der Wahrnehmung auftreten, habe ich im Vorigen gegenübergestellt. Zusammenfassend sei gesagt, dass erstens bedeutet es das *Transzendieren des wahrgenommenen Dinges* über jede einzelne Wahrnehmung, über das in erfüllten Qualitäten Gegebene hinaus. Es wird in der Wahrnehmung gegeben, dass das Ding ein Inneres und eine Hinterseite hat, aber in der betreffenden Wahrnehmung ist das nicht in erfüllten Qualitäten gegeben, sondern es ist irgendeine anschauliche Intention darauf vorhanden, dass man die Hinterseite hat, dass man ein Inneres hat, usw. *Zweitens* bedeutet es das *Transzendieren des Wahrnehmens* über diese erfüllten Qualitäten des wahrgenommenen Dinges mit den Vermeintlichkeiten auf das Innere. Einerseits gibt es eine Transzendenz von dem Ding aus, und das andere ist eine Transzendenz vom Wahrnehmen aus. Der dritte Begriff der Transzendenz, auch von dem epistemologischen, erkenntnisstheoretischen Standpunkt, ist er die Transzendenz des physikalischen Dinges, des Dinges, das so verstanden wird, wie es die moderne Physik tut.[3] Diese Transzendenz des physikalischen Dinges ist aber Transzendenz eines erkennbaren Dinges, und zwar erkennbar durch die Zurückführung der Hypothese auf die Erfahrung; es ist nicht transzendent in einem noch anderen Sinne, von dem ich gleich sprechen werde (von der Transzendenz des Unerkennbaren). Es ist dabei hervorzuheben dass diese theoretisch gebildete Welt, die mathematisch bestimmte Welt, eine gewisse merkwürdige "Reperkussion" auf die von uns wahrgenommene Welt hat; sie ermöglicht eine Interpretation von Phänomenen, von Gegebenheiten im Sinne der Theorie. Man kann eine Menge von solchen Tatsachen angeben, wo in der Erfahrung etwas gegeben wird, was ohne die Theorie überhaupt nicht beachtet und auch nicht verstanden sein würde, man interpretiert es als im physikalischen Sinne seiend z.B. Interferenz, Interferenzstreifen. Da hat man z.B. auf der Leinwand (*écran*) bei bestimmter Einrichtung des Experiments (das kann auf verschiedene Weise gemacht werden, mit Spiegeln) jene merkwürdigen hellen und dunklen Streifen. Der Physiker deutet es als eine Wellenkreuzung, ein Zeichen davon, dass da zwei Wellenpakete sich kreuzen und an einzelnen Stellen entsteht Stille. Der Stille entspricht das Dunkel und der verstärkten Bewegung entspricht das Helle. Ein anderes Beispiel bilden die sogenannten Fraunhoferschen Linien im Sonnenspektrum. Wenn man so ein Spektrum durch ein besonderes Glas (das Prisma) durchführt, bekommt man alle möglichen Farben; und dann zeigt es sich, dass es dabei verschiedene dunkel Streifen gibt. Das beruht auf der Theorie, welche diese Streifen auf eine ganz ingeniöse Weise interpretiert. Wenn zwischen dem Objekt, welches Licht emittiert, und dem Apparat, der es aufnimmt, ein ganz bestimmtes Gas vorhanden ist, welches die entsprechenden Lichtwellen verschluckt, dann herrscht im aufnehmenden Apparat an entsprechender Stelle Stille, also eine dunkel Linie. Infolgedessen interpretiert man dann die Fraunhoferschen Linien am Sonnenspektrum sofort als ein sichtbares Zeichen des Vorhandenseins z.B. (in der Sonnenkorone) solcher und nicht anderer Gase. Die Theorie mit entsprechenden rechnerischen Mitteln ermöglicht nicht nur eine genaue Berechnung der Stellen dieser Linien, sondern trägt dazu bei, dass das Gegebene in der Wahrnehmung sofort auf eine bestimmte Weise gedeutet und verstanden wird.[4] Dann z.B. die für uns jetzt sehr wichtige "Rotverschiebung". Wenn man das Licht von grossen Galaktiken aufnimmt, die sehr weit von uns entfernt sind, dann zeigt es sich, dass an ihrem Spektrum die Fraunhoferschen Linien anders liegen als sie nach der Theorie liegen sollten, sie werden stark nach rechts verschoben. Dies wird mit Berufung auf das Dopplergesetz so interpretiert, dass die Rotverschiebung das "signum" davon ist, dass die betreffende Galaktik sich sehr schnell von uns fortbewegt.

Ihre Fortbewegung ist ihrer Entfernung von der Erde proportional, je weiter desto schneller. Man macht Aufnahmen davon, wo solche Galaktiken ein paar Milliarden Lichtjahre von uns entfernt sind und die Geschwindigkeit schon 80% der Lichtgeschwindigkeit ausmacht. Die H- und K-Linien verschieben sich hier sehr weit nach rechts. Der Physiker behauptet: diese Tatsachen bedeuten dass dies in der Wahrlichkeit geschieht. Die Rotverschiebung ist nicht einfach eine Tatsache für sich, sondern soll ein phänomenales Zeichen von etwas ganz Anderem sein. Die Tatsachen, die in der Wahrnehmung beim Experiment gegeben werden, werden sofort als Erscheinung einer Tatsache in der physikalischen Welt gedeutet.

Duhem, ein französischer Erkenntnistheoretiker, behauptet, dass jede Tatsache in der Naturwissenschaft schon Ergebnis einer Theorie ist. Eine "Tatsache" ist nicht einfach dasjenige, was schlicht ohne die Theorie wahrgenommen wird. Ohne die Theorie gibt es gar keine naturwissenschaftlichen Tatsachen. Erst wenn man die Theorie der Fraunhoferschen Linien und das Doppler-Gesetz kennt und in diesem Sinne die Rotverschiebung im Lichtspektrum einer Galaktik deutet, ist die Rotverschiebung eine "Tatsache". Wenn man die sogenannten kosmischen Strahlen untersucht (das sind Partikelchen, die von der Welt auf die Erdoberfläche fallen und alles mögliche durchdringen), entstehen analoge "Tatsachen". Man sendet Päckchen von photographischen Platten (etwa 20 km) in die Luft; und zieht man sie herunter, so zeigt es sich, dass sich an diesen Platten kleine schwarze "Sternchen" befinden. Dies wird als ein Ereignis interpretiert: die lichtempfindliche Substanz der Platte wurde verändert. Man betrachtet diese Sternchen durch das Mikroskop und sieht die sich kreuzenden Linien als Spuren der Bewegung eines Partikelchens. Das "Sternchen" ist ein Hinweis auf eine Katastrophe in der Atomenwelt. Ein Atomkern wurde durch ein Alpha-Teilchen oder durch einen Meson zerschlagen und seine Teile fliegen weiter nach verschiedenen Richtungen und die Spuren ihrer Wege sind auf der Platte angezeichnet. Aus den Winkeln zwischen den Linien in einem solchen Sternchen kann man die Energie der Teilchen vor und nach der Kollision berechnen. Daraus ist eine neue Theorie der "Elementarteilchen" entstanden. Ein solches "Sternchen" ist die reine, ohne die Theorie wahrgenommene Tatsache. Für einen Physiker ist das Sternchen nicht einfach eine Zeichnung auf der Platte, sondern die Spur einer Katastrophe in der Atomenwelt. Man kann daraus lesen was für eine Katastrophe sich abgespielt hat; man kann auch fragen, was für ein Atom vor der Katastrophe vorhanden war und was mit ihm geschah. Dies alles kann man berechnen. Nach der Berechnung, nach dieser Interpretation, wird das "Sternchen" zu einer physikalischen Tatsache; es wurde im Phänomen sichtbar, aber musste durch die Theorie interpretiert werden.

Das ist also die "Transzendenz" des physikalischen Dinges in Bezug auf die Erfahrung, auf eine Mannigfaltigkeit von verschiedenen, im Zusammenhang stehenden Wahrnehmungen. Die physikalische Tatsache (Ding, Vorgang, Ereignis) wird *nie* selbst gegeben; ihr Vorhandensein wird in dem uns mittelbar Gegebenen angezeigt, jedoch nicht ohne Hilfe einer Theorie.

Wir kommen jetzt zu dem *vierten* erkenntnistheoretischen Begriff der Transzendenz. Diesen Begriff verdanken wir Kant, und zwar seiner *Kritik der reinen Vernunft*. Da wird zwischen Erscheinungswelt und Welt der Dinge an sich unterschieden. Die Erscheinungswelt ist die Welt der wahrgenommenen Dinge. Eine "Erscheinung" ist nicht – wie man vielleicht heute annehmen könnte – eine Ansicht, oder eine Wahrnehmung, es ist das wahrgenommene Ding, z.B. das gesehene Glas Wasser; und das Ding an sich ist das, was *überhaupt* nicht erkennbar ist. Man fragt sich warum ist es für Kant nicht erkennbar? Wohl, weil subjektive und zwar transzendentalsubjektive Faktoren mitwirken, nämlich die apriori'schen Anschauungsformen Raum und Zeit und das System der Kategorien. Das Ergebnis der Kant'schen Untersuchung liegt in der Behauptung, dass die apriori'schen Anschauungsformen und die reinen Verstandesbegriffe uns eine objektive Erkenntnis ermöglichen, weil wir alle dieselben Kategorien und dieselben Anschauungsformen besitzen. Dadurch haben wir eine gemeinsame Erscheinungswelt. Die Anwendung dieser apriori'schen Formen bringt aber andererseits eine Transformation, eine phänomenale Transformation der Wirklichkeit mit sich; wir sind dadurch von der Welt der Dinge an sich abgeschnitten.

Wie ich erwähnt habe, gibt es verschiedene Interpretationen der Kant'schen Transzendentalen Philosophie. Die eine besagt, das System der Kategorien und die Anschauungsformen seien für jedes mögliche Erkenntnissubjekt wesensnotwendig. Heidegger hat diese Interpretation eingeengt, indem er hinzugefügt hat: "für jedes *endliche* Erkenntnissubjekt", das er dem Intellektusarchitypus d.h. Gott gegenübergestellt hat, der nicht gezwungen ist, die Welt in diesen Strukturen zu erkennen. Wenn aber der Kant'sche Standpunkt nicht auf die Heidegger'sche Weise gedeutet wird, dann ist das Ding an sich, ganz radikal gesprochen, für kein Erkenntnissubjekt erkennbar, dann also auch nicht für Gott. Endlich gibt es noch eine dritte Interpretation, nämlich die anthropologische Interpretation der *Kritik der reinen Vernunft*. Nach ihr sind die "apriori'schen" Strukturen, die uns von der Welt der Dinge abschneiden, nur auf uns Menschen, beschränkt. Wir Menschen haben so eine besondere subjektive Natur, dass wir immerfort Anschauungsformen wie Raum und Zeit und Kategorien verwenden müssen. Dann sind die Dinge nicht allgemein, sondern nur für uns Menschen unerkennbar. Was aber die Dinge an sich betrifft, so lassen sich darüber verschiedene Interpretationen des Kant'schen Standpunktes angeben. Entweder sollen die Dinge an sich weder räumlich noch zeitlich sein, noch in Kategorien stehen, so dass ihre apriori'sche Formung nur für die menschlichen Subjekte gilt, oder es darf nur behauptet werden, dass die apriori'sche Formung uns jeglichen Zugang zu den Dingen verschliesst, so dass wir gar nicht wissen können, ob sie wirklich raum- und zeitlos und kategorienlos sind. – Dieser Kant'sche Begriff der Transzendenz prägt die europäische Philosophie seit dem 18. Jahrhundert und es wurden viele Versuche unternommen, diesen Begriff und die mit ihm zusammenhängende Scheidung zwischen den Dingen an sich und den Erscheinungen irgendwie zu überwinden. Die deutschen Philosophen-Romantiker haben es auf verschiedene Weise versucht. Sowohl Schelling als auch Fichte, und Hegel, und endlich Schopenhauer, jeder von ihnen auf eine eigene Weise, suchten Zugang zu den Dingen an sich zu gewinnen. Die Scheidewand zwischen uns und der Wirklichkeit wollte man irgendwie durchbohren, und jeder von ihnen hat das auf seine eigene Weise getan. Man kann sagen, dass Kant selbst diesen Versuch in der *Kritik der praktischen Vernunft* unternommen hat. Bergson hat es auf eine wiederum andere Weise getan – etwas anklingend an Schelling. Ganz neu war die Relativierung der Kategorien auf den Intellekt und die Erfordernisse der Handlung. Die Neokantianer, und insbesondere die Marburgerschule vertrat wie bekannt die Auffassung, dass das Ding an sich überhaupt zu streichen ist. Denn wenn Anschauungs- und Verstandes-Formen wirklich notwendige apriori'sche Formen sind und zum Wesen des Erkenntnissubjektes gehören, so dass jedes Subjekt alles überhaupt nur in diesen Formen erkennen kann und soll, dann ist das Ding an sich absolut unerkennbar und als solches einfach zu streichen. Damit verwandelt sich die ganze Erkenntnis in eine unendliche Aufgabe, immer neue Schritte im Erkennen zu vollziehen, das Gegebene immer als für neue Formung im unendlichen Fortgang aufgegeben zu halten. Husserl stand Kant zunächst im Sinne der Tradition von Franz Brentano sehr fern und kritisch gegenüber. Brentano war bekanntlich gegenüber der ganzen deutschen Philosophie, und speziell gegenüber Kant, feindlich gesinnt, und das hat Husserl gewissermassen übernommen. Aber es hat sich allmählich gezeigt – bereits in den *Ideen zu einer reinen Phänomenologie* – dass es zwischen ihm und Kant gewisse Anknüpfungspunkte gibt, und zwar auch zu dem Marburger, Paul Natorp. In Natorps *Allgemeine Psychologie* gibt es gewisse Berührungspunkte zu den *Ideen* von Husserl. Der Punkt, auf den es mir jetzt ankommt, bildet die Stellung der beiden Philosophen zu dem Begriffe des Dinges an sich. Das formuliert Husserl folgendermassen: Das Ding an sich oder überhaupt jedes Seiende, das unerkennbar ist und zwar prinzipiell, das heisst, wenn es zu seinem Wesen gehören sollte, dass es unerkennbar ist, ist ein Widersinn und darf als solches nicht angenommen werden. Der gemeinsame Punkt Husserls mit Paul Natorp und mit der Marburgerschule liegt darin, dass man den Begriff des Dinges an sich, der Transzendenz im Sinne des absolut Unerkennbaren, verwirft.

In diesem Zusammenhang ist der Begriff von "Scheinproblem" Carnaps zu erwähnen. Dies sind eben Probleme, die sich auf das Unerkennbare beziehen. Nach Carnap sind dies von Grund auf falsch gestellte Fragen, die man überhaupt nicht sachlich behandeln soll. Man hat damals, als das Büchlein erschien, es bei den Neopositivisten als grosse Neuentdeckung gerühmt. Tatsächlich aber spricht diesen Gedanken schon Husserl aus, nur dass der Bereich der Probleme, die Carnap als "Scheinprobleme" verwirft, von Husserl ganz anders abgegrenzt ist.[5]

Kehren wir aber zu Husserl selbst zurück. Er behauptet: "Das Unerkennbare streichen wir, darüber sprechen wir nicht". Daraus ergibt sich, wenn etwas als Seiend anerkannt werden soll, muss es einen Weg zu seiner Erkenntnis geben. Jede Erkenntnis oder jeder Erkennende bleibt in einer Welt der Phänomene. Das ist der erste und vielleicht auch der weiteste Sinn der Phänomenologie; von Dingen an sich sprechen wir nicht, weil es Widersinn ist; und das, worüber wir sprechen, was wir erkennen wollen und können, das sind "Phänomene", das ist dasjenige, was dem Erkennenden irgendwie phänomenal zugänglich ist. Dadurch wird der erste Schritt zu dem getan, was sich später als transzendentaler Idealismus entwickelt.

Um Husserls Position weiter zu erhellen werde ich jetzt auf Husserls Vorlesungen Bezug nehmen. Zwischen 1912 und 1914 hat Husserl die folgenden Ideen vorgetragen: Husserl stellte damals zwei Sätze nebeneinander, nämlich: (1) "*A* existiert", und (2) nach wörtlicher Formulierung Husserls: "Es ist ein Weg zur Ausweisung von *A* prinzipiell möglich". Diese beiden Sätze sollen nach Husserl äquivalent sein. Da fällt das Wort "Ausweisung" auf. Was soll dieser Begriff bedeuten? Wichtig ist hier auch die Wendung "ein Weg zu etwas prinzipiell möglich". Das Wort "Ausweisung" ist bei Husserl ein *terminus technicus*, es bedeutet, dass, wenn man irgendwelche Begriffe, Vermeinungen oder Behauptungen über etwas hat, man in der unmittelbaren Erkenntnis zeigen muss, dass es stimmt. Wenn man z.B. von diesem Ding spricht und seine verschiedenen Eigenschaften feststellen will, dann muss man in der Wahrnehmung zeigen, dass es so ist. "In der Wahrnehmung" das soll nicht besagen "eine Wahrnehmung". Es handelt sich gewöhnlich um eine ganze Mannigfaltigkeit von Wahrnehmungen, die das betreffende Ding von verschiedenen Standpunkten aus, von verschiedenen Seiten und unter verschiedenen Umständen so hervorhebt, dass der behauptete Tatbestand als wirklich bestehend gezeigt wird. Wenn man diesen ganzen Prozess des Sichzeigens des betreffenden Gegenstandes oder seiner Eigenschaft in den Mannigfaltigkeiten von Wahrnehmungen durchlaufen hat und alles miteinander abstimmt, dann sagt man die Behauptung sei "ausgewiesen", es sei insbesondere ausgewiesen, dass der Gegenstand existiert. Die Wendung "ein Weg ist prinzipiell möglich" soll besagen, dass nicht nur wir, Menschen (und eventuell auch Tiere), sondern überhaupt *jedes* nur mögliche Erkenntnissubjekt, wie es auch immer gebaut sein und welche Fähigkeiten es sonst haben mag, einen Erkenntniszugang zu dem Gegenstand *A* besitzen muss, damit *A* in seiner Existenz ausgewiesen werde. Die Existenz bzw. die prinzipielle Möglichkeit dieses Weges ist die Bedingung der Behauptung, dass *A* existiert, oder schärfer gesagt: dass das Recht, die Anerkennung der Existenz von *A* besteht. Aber so lauten die beiden zitierten Sätze nicht. Es ist gesagt, es sei der prinzipiell mögliche Weg zur Ausweisung von *A*, mit dem "*A* existiert" äquivalent. Das heisst genauer, der Weg zur Ausweisung ist äquivalent mit der Existenz.

Wo liegen aber die Begründungen der Möglichkeit, *A* auszuweisen? Man kann sagen: beiderseits, an *A*, am Gegenstand, und an dem Erkenntnissubjekt. Das Existierende muss in sich selbst so beschaffen sein, dass es sich dem Erkennenden zeigen oder sich wenigstens vermittels eines anderen Gegenstandes, der diese Fähigkeit besitzt, anzeigen kann. Der Gegenstand, der direct ausgewiesen werden soll, muss fähig sein, im Phänomen für ein mögliches Erkenntnissubjekt zu erscheinen, ein Phänomen, von dem aus erst seine anderen Eigenschaften entdeckt werden können. Wenn man "Phänomen" sagt, das heisst Phänomen für etwas, für jemanden, "Phänomen für". Sartre würde sagen "c'est être pour quelqu'un".

Im Sein selbst muss diese Fähigkeit enthalten sein, dass es für jemanden erscheinen könnte, dass es diese Qualität in sich hat. Oder noch mehr, sein Sein *ist* so erkennbar. Und das ist die Bedingung, dass es existiert, dass es daneben noch den erkennenden Geist geben muss. Sein Sein ist in seinem Sein durch das Bewusstsein bedingt, in dem es zur Ausweisung gelangt. Es gehört zum Sinn der Existenz auf Grund dieser Äquivalenz, dass das Existierende ein (mögliches) Korrelat zu einem Bewusstsein ist. Und wenn es ausgewiesen werden soll, dann muss es ein wirkliches Korrelat eines Bewusstseins sein. Das ist das Gegenteil zu dem Kant'schen Begriff des Dinges an sich, da es zum Wesen des Dinges gehört, dass es von einem möglichen Erkenntnissubjekt nicht mit Bewusstsein erfassbar ist; dass es jede mögliche Erfassungsweise absolut transzendiert. Bei Husserl ist es umgekehrt: es gehört zum Wesen des existierenden Dinges, des existierenden "Etwas", dass es Korrelat zu einem Bewusstsein ist oder mindestens sein kann. Natürlich gibt es Bedingungen von der anderen Seite. Das Bewusstsein, das Erkennende, muss von sich aus solche Operationen oder solche Bewusstseinsverläufe haben, dass es einen Zugang zu dem seienden *A*, zum seienden Gegenstand, gewinnt. Wenn man dagegen das Erkenntnissubjekt von vornherein so fasst, wie es Kant getan hat, dann – würde Husserl sagen – ist das der Grundfehler der ganzen Theorie. Denn da wird das Subjekt so konzipiert, so gestaltet, dass es durch die apriori'schen Kategorien und die Anschauungsformen eine Mauer zwischen sich und der Wirklichkeit aufwirft.

Es wird aber zunächst nicht gesagt, wie es möglich ist, zu den Dingen erkenntnismässigen Zugang zu gewinnen. Es muss jetzt untersucht werden, ob dieses Erkenntnissubjekt nich zufälligerweise gerade solche Erlebnisverläufe hat, dass es sich von der Wirklichkeit abschliesst. Dann kämen wir zu Kant zurück. Und da erhebt sich ein grosses Problem. Es müssen subjektiv orientierte Analysen durchgeführt werden: wie alles verläuft, damit es in diesen Verläufen zu Phänomenen kommt, die uns das Ding selbst, die Wirklichkeit, wie sie ist, um das Heidegger'sche Wort zu benützen, "enthüllen", entdecken könnten. Die Verwerfung des Kantschen Transzendenzbegriffes bringt uns jetzt an den Rand der transzendentalen Betrachtung, aber zugleich auch an den Rand des transzendentalen Idealismus. Diese Entscheidung steht uns jetzt ganz nahe.

Soviel über die verschiedenen Transzendenzbegriffe, die *erkenntnistheoretisch* formuliert sind, mit Bezug auf die mögliche oder wirkliche Erkenntnis des Erkenntnisgegenstandes.

Daneben gibt es aber noch eine *zweite Gruppe* von Transzendenzen, die noch zu besprechen sind: Diejenigen Begriffe die Husserl im Zusammenhang mit dem Problem der Beziehung zwischen dem Bewusstsein von etwas und den physischen Dingen positiv bestimmt. Die Transzendenz wird jetzt ganz anders gefasst. "Bewusstsein" das sind vor allem Erkenntniserlebnisse. Es werden aber nicht nur diese Erlebnisse sondern überhaupt *jedes* Bewusstsein in Betracht gezogen. Die Beziehung zwischen materiellem Ding und Bewusstsein wird jetzt ontisch aufgefasst. Wie dieses geschieht werde ich aus dem Text der *Ideen* lesen.[6]

> Nun könne ein Wahrgenommenes selbst sehr wohl Bewusstseinserlebnis sein; aber es ist evident, dass so etwas wie ein materielles Ding, z.B. dieses im Wahrnehmungserlebnis gegebene Papier, prinzipiell kein Erlebnis ist, sondern ein Seiendes von total verschiedener Seinsart, im Erlebnis als real seiend bewusst, aber nicht als reales Bestandstück darin enthalten.[7]

Also es ist kein Erlebnis, es ist ein Seiendes von total anderer Seinsart, und ist kein "reales Bestandtstück" des Bewusstseins, des Erlebnisses. Husserl sagt übrigens gewöhnlich nicht "reales", sondern "reelles" Bestandstück, reelles Stück des Erlebnisses. Diesen Ausdruck ("reelles Bestandstück") hat er schon in den *Logischen Untersuchungen* bei der Analyse der Bewusstseinsakte, der intentionalen Akte verwendet. Zwischen "real" und "reell", wie Grüber bemerkt, besteht ein wesentlicher Unterschied. "Real" ist ungefähr dasselbe wie "wirklich"; zu der "reellen" Welt gehörig dagegen bedeutet soviel wie ein "wirklich echter" Teil, "effektiver" Teil sein. Reelles Bestandstück, reeller Teil, das bedeutet soviel wie im "echten Sinne Teil von etwas zu sein". Wenn also

Husserl sagt, das physische (materielle) Ding sei kein "reelles Bestandstück" des Bewusstseins, so heisst das, dass es gar kein effektiver, echter Teil vom Bewusstsein ist, obwohl es im Bewusstsein erfasst wird.

Es sind da zunächst zwei Punkte hervorzuheben. Der erste Punkt dieses neuen Begriffes der Transzendenz bezieht sich auf die *Seinsart*, auf die Seinsweise des materiellen Dinges. Das gehört in meiner Sprache zur "Existenzial-Ontologie". Und der zweite Punkt bezieht sich darauf, dass dieses Ding *ausserhalb* des Bewusstseins existiert. Es ist etwas Zweites im Vergleich zum Bewusstsein. In meiner Sprache ist das ein "formales Moment". Es liegt in der Form des Dinges und des Erlebnisses, dass es das Eine und das Zweite ist. Und jetzt, dieses Zweite, dieses Nicht-reeller-Teil-sein-von", kann noch auf zwei verschiedene Arten verstanden werden, worauf ich gleich zurückkomme.

Zu dem ersten Punkt dieses Transzendenzbegriffes ist zu fragen, in welchem Sinne es eine andere Seinsweise, die Seinsart des materialen Dinges ist. Eine "andere" wovon? Von dem Sein, von der Seinsweise des Bewusstseins, des Erlebnisses? Zunächst ist zu bemerken: Wenn man dieses liest, so weiss man nicht, wie diese andere Seinsweise zu verstehen ist. Im Laufe der ganzen Betrachtung aber klärt sie dies, indem es zunächst auf etwas zurückgeführt wird, was wir jetzt noch weiter zu behandeln haben.

Bei Husserl stehen zwei Begriffe sehr nahe aneinander und werden fast äquivalent benutzt, das ist: "Seinsweise" und "Gegebenheitsweise", oder "Erkenntnisweise". Husserl behauptet, dass Dinge durch Abschattungen, durch Ansichten, durch Erscheinungen wahrgenommen werden. Dagegen Erlebnisse werden nicht durch Abschattungen erfasst, sondern in einer immanenten Wahrnehmung. Sie sind ohne Abschattung. Es gibt gar keine Abschattung, keine Ansicht von meinem Erlebnis, sondern es wird selbst, sozusagen, ohne Hilfe oder Vermittlung der Abschattung erfasst.

Husserl unterscheidet nun die immanente Wahrnehmung als eine andere Erscheinungsweise, andere Erkenntnis Weise des Dinges und meines Bewusstseins. Es wird später von Husserl gezeigt, in welchen Abschattungen Dinge erscheinen und wie das eigenene Bewusstsein, das eigene Erlebnis erfasst wird. Nun aber schliesst Husserl sofort: "mein in der Immanenz erfasstes Bewusstsein existiert absolut, und das in Abschattungen, in Erscheinungen, erfasste materielle Ding existiert nicht absolut". Die

Frage entsteht, wie es denn existiert. Am Schluss der ganzen Betrachtung wird von Husserl behauptet es sei ein "intentionales Korrelat". Das materielle Ding existiert nicht absolut für sich selbst als intentionales Korrelat von Bewusstseinsmannigfaltigkeit. Wenn wir jetzt auf die zwei früher erwähnten Sätze zurückkommen: "*A* existiert", das heisst, "es gibt einen Weg zur Ausweisung des *A*". Das existierende *A* ist nur als das zu nehmen, was prinzipiell ausweisbar ist, was als Korrelat des Ausweisungsbewusstseins besteht. Das Sein von *A* ist bedingt durch das Sein des Bewusstseins, es ist dessen Korrelat. Nur als Korrelat existiert es und ausschliesslich auf diese Weise. Ohne die oben dargestellte Theorie der Wahrnehmung (dass Dinge in Mannigfaltigkeiten von Erscheinungen erfasst werden, wahrgenommen werden – von Erscheinungen, von Ansichten, von Abschattungen, das sind die drei Ausdrücke, die Husserl verwendet); würde man – wenn man sich genau an die Texte Husserls hält – auch nicht sagen können, dass sie anders auf andere Weise als das Wahrnehmungserlebnis existieren. Das einzige Argument bei Husserl für diese These kommt aus der Erkenntnis dass sie anders erkannt werden, folglich existieren sie anders.

Bei dieser Einführung des *ontischen* Begriffes der *Transzendenz*, sieht man diese Hintergründe nicht – man fragt sich: Wo kommt diese zweite Seinsart her? Denn der zweite Punkt – das formalontologische Moment der Bestimmung der Transzendenz des materiellen Dinges dass es kein reeller Teil des Bewusstseins ist – lässt doch gar nicht folgen dass es ein Sein einer anderen Art ist; dass das Ding eine andere Seinsweise hat. Es wird nur formal zwischen den beiden Intentionskorrelaten unterschieden. Es muss also noch andere ergänzende Gründe zur Bestimmung der Transzendenz des materiellen Dinges geben. Sie sind bei Husserl eben epistemologischer Art.

Jetzt muss ich noch die zwei Sinne dieses "Auseinanderseins" genauer erklären. Denn Husserl äussert sich einmal[8]:

> Die Wahrnehmung des Dinges enthält nicht nur in seinem reellen Bestande das Ding selbst nicht in sich, es ist auch ausser aller wesentlicher Einheit mit ihm, seine Existenz natürlich vorausgesetzt. Eine rein durch die eigenen (singulären)[9] Wesen der Erlebnisse bestimmte Einheit ist ausschliesslich die Einheit des Erlebnisstromes, oder was dasselbe ist, ein Erlebnis kann nur mit Erlebnissen zu einem Ganzen verbunden sein, dessen Gesamtwesen die absoluten eigenen Wesen dieses Erlebnisses umschliesst und in ihnen fundiert ist.

Anders ausgedrückt, es würde sich um die Interpretation des "nicht-reelles-Bestandstück-vom-Bewusstsein-zu-sein" handeln. Wir haben nämlich zwei Gegenständlichkeiten, einerseits das Bewusstsein, und andererseits das Ding. Es sind zwei Gegenständlichkeiten, aber es sind nicht nur auf diese Weise zwei, dass sie überhaupt unterschieden werden können, wie z.B. die Farbe und die Gestalt einer Kugel, sondern es sind zwei Ganzheiten. Jede davon ist eine Ganzheit für sich; sie sind in gewissem Sinne gegenseitig abgeschlossen. Weiterhin gibt es keine höhere Struktur, die da beides zu *einem* Ganzen umfassen würde. Es sind zwei Ganzheiten, die ausserhalb voneinander bestehen. Wo liegt dieses doppelte Auseinandersein, oder besser, dieses Aussereinandersein von zwei Ganzheiten, die nicht miteinander verbunden sind, nicht vor? Husserl findet es ausschliesslich bei Erlebnissen. Wenn ich z.B. jetzt nach links sehe und zugleich über etwas nachdenke, vollziehe ich eine Mannigfaltigkeit von verschiedenen Erlebnissen; ich habe mir etwas vorgestellt, ich fühle meinen Leib, usw., und alle diese Erlebnisse fügen sich zu einem Ganzen zusammen. Sie bilden *ein* Ganzes, eine Phase meines konkreten bewussten Lebens. Sie heben sich freilich in ihm ab, sind aber doch in dem Sinne voneinander unselbständig, dass sie aus diesem konkreten Erlebnisganzen nicht herausgelöst werden können. Dagegen in dem Fall, wo wir es einerseits mit einem Wahrnehmungserlebnis von einem physischen Ding und andererseits mit diesem Ding selbst zu tun haben, da sind es zwei seinsmässig und auch in ihrem Beschaffensein gegenseitig voneinander selbständige Ganzheiten.

Ein derartiges nicht-Teil-von-Etwas-sein bildet die ontische Transzendenz des materiellen Dinges den Bewusstseinsakten, den Erlebnissen gegenüber. Es könnte auch anders sein. Es könnte z.B. so sein, wie in dem Falle, wo ein materielles Ding vorhanden ist, das farbig ist und zugleich glatt und zugleich kalt, also eine bestimmte Temperatur hat. In diesem einen Ding gibt es drei verschiedene Momente, drei verschiedene Eigenschaften die auch getrennt bestehen; Farbe, Temperatur und glatte Oberfläche. Oder z.B. ein Klang, den dieses Ding gerade hervorbringt, der so oder anders klingt. Es sind alles unselbständige Momente *dieses* einen Dinges; ohne dieses Ding können sie nicht existieren. Sie sind irgendwie miteinander verflochten, verschwommen, jedoch sie sind von einander verschieden. Ausser der Farbe gibt es den Klang des Dinges, und die Temperatur, aber trotzdem sind sie in dem Sinne voneinander unselbständig, dass diese konkreten Bestandteile des Dinges voneinander nicht zu trennen sind. Sie können vielleicht auf eine bestimmte Weise irgendwie variieren, aber sie verbleiben doch unselbständige Momente desselben Dinges.

Husserl aber behauptet dass mit der (ontischen) Transzendenz des Dinges in Bezug auf das Erlebnis, in dem es wahrgenommen wird, und überhaupt in Bezug auf jedes Erlebnis, dies nicht der Fall ist.

Nach ihm gibt er keine Einheit, keine sich aus dem Wesen der beiden Elemente ergebende Form, die sich zu *einem* innerlich verbundenen Ganzen vereinheitlichen würde. In diesem Sinne ist das materielle Ding dem Erlebnis gegenüber "transzendent". Das eine ist eine "cogitatio", sagt Husserl später, und das andere ist "cogitatum", es sind zwei verschiedene Gegenständlichkeiten. Es gibt aber Fälle, wo dies nicht zutrifft. Nehmen wir die immanente Wahrnehmung meiner transzendenten äusseren Wahrnehmung eines Dinges in Betracht, oder, wenn ich jetzt denke und darauf achte, wie sich dieser Prozess vollzieht, dass ich denke und nicht stehen bleibe, sondern weiter denke, obwohl ich gleichzeitig Dinge im Saale sehe, dann scheint es als ob ich im gewissen Sinne doch immanent wahrnehme. In diesem Falle sagt Husserl dass das "cogitatum" zu demselben Erlebnisstrom wie das "cogitare" gehört, meine "cogitatio", meine immanente Wahrnehmung, und das immanent Wahrgenommene, mein Denken und mein Wahrnehmen gehören jetzt zu demselben Bewusstseinsstrom. Und dazu, wenn das wirklich eine echte immanente Wahrnehmung ist, dann liegt nach Husserl noch etwas ganz Merkwürdiges vor: Nämlich, immanent Wahrgenommenes, und das immanente Wahrnehmen bilden eine unvermittelte Einheit, die Einheit einer einzigen konkreten "cogitatio". Sie bilden *ein* Ganzes, und zwar so, dass das Wahrgenommene, – das heisst mein jetzt sich vollziehendes Denken – und das sich-zum-Bewusstsein-bringen dieses Denkens in dem Sinne eine Einheit sind, d.h. dass das Wahrgenommene in dem Wahrnehmen "reell beschlossen" ist.[10]

Das wahrgenommene Denken bildet die Basis auf der sich das immanente Wahrnehmen meines Denkens aufbaut. Es sind zwei noch unterscheidbare "Gegenständlichkeiten", aber trotzdem ist das Wesen des immanenten Wahrnehmens so, dass es ohne das in ihm Wahrgenommenen nicht bestehen könnte. Das Wahrnehmen meines Denkens, dieses individuelle Wahrnehmen, kann nicht existieren ohne das Denken, das da gerade wahrgenommen wird; es ist in seinem Sein auf dieses Denken relativ unselbständig. Wenn das Wahrnehmen existiert, dann existiert natürlich das Wahrgenommene, das heisst mein aktuelles Denken.

Bei Descartes, "cogito – sum", sieht man dasselbe, bloss auf das "sum" bezogen. Aber das "cogito" existiert; wenn ich es feststelle, und wenn ich das "cogito" erfasse, dann existiert auch das "cogitatum". Mein "cogito" und das "cogito" des "cogito", das ist eine Einheit und in diesem Fall ist es aus dem Wesen der "cogitationes" nicht nur möglich aber auch notwendig: Sie sind eins. Man kann also sagen, wenn es das Selbsterfassen, die Selbstwahrnehmung – also die immanente Wahrnehmung – gibt, dann verbürgt mir die Selbstwahrnehmung die Existenz des Wahrgenommenen, meines Erlebnisses, weil gerade das Erste nicht ohne das Zweite existieren kann. Es ist ihm gegenüber unselbständig. Dieselbe Einheit aber kann es zwischen Ding und dem Wahrnehmenden wesensmässig nicht geben, da sie auseinander sind und auf eine solche Weise getrennt sind, dass sie keine neue Ganzheit bilden können.

Wir besitzen also jetzt die Begriffe der (ontischen) Transzendenz und der Immanenz bei Husserl. "Und doch ist es klar" – führt Husserl unmittelbar weiter, ... dass die "cogitatio" in sich "cogitatio" von ihrem "cogitatum" ist, und dass dieses als solches, so wie es da ist, von ihr unabtrennbar ist", "das *es* das "cogitatum" ist". Wovon ist "cogitatum" untrennbar? Die Antwort drängt sich auf: Von *ihr*, d.h. von der "cogitatio". Das Ding ist transzendent, das Ding ist in diesem Falle "cogitatum" Es ist transzendent, d.h. es bildet kein Stück von dem Wahrnehmen und auch steht mit ihm ohne jede wesentliche Einheit, so dass es kein Ganzes mit ihm bildet, und doch, es ist von ihm unabtrennbar. Soll mann darunter verstehen, dass was "cogitatum" ist, ist unabtrennbar vom "cogitare"? Es ist dabei zu beachten, dass nicht gesagt wird, dass "cogitare" unabtrennbar vom "cogitatum" sei, sondern umgekehrt. Es ist zwar ein Ganzes, ein zweites, ohne wesentliche Einheit mit ihm, jedoch "unabtrennbar" von ihm. Was aber dieses "unabtrennbar" heisst, wird zunächst nicht gesagt. Später klärt es sich und zwar in dem Sinne, dass es nicht bestehen kann ohne dass sich die betreffende "cogitatio" vollzieht. Wir lesen diesen Satz in den *Ideen* ohne eine nähere Begründung zu bekommen. Wir müssen uns fragen, was die Gründe dafür sind, dass dieses "cogitatum" das Transzendente sein soll, und in keiner Einheit mit dem "cogitare" steht, und doch von diesem "cogitare" unabtrennbar ist.[11]

Wir kommen zurück zu der Äquivalenz der Sätze "*A* existiert" und "es ist ein Weg prinzipiell möglich, die Existenz des *A* auszuweisen". Hier muss man sagen: Obwohl es Zweites und Transzendentes ist, existiert es auf die Weise, dass es ohne das "cogitare" nicht existieren kann. Das bildet das Wesen dieser Existenzweise. Es ist eben, behauptet Husserl, kein absolutes Sein; dagegen ist das "cogitare" der "cogitatio" ein absolutes Sein, das existieren kann ohne dass das existierende Ding da ist. Es kann existieren, da es durch das Wahrgenommene nicht bedingt ist.

Das ist die asymmetrische Beziehung zwischen zwei Seienden; das erste Seiende, das ist das Bewusstsein, das Erlebnis, und das zweite Seiende, das ist das materielle Ding und die ganze materielle Welt. Diese Beziehung geht noch weiter. Auch ich als Mensch bin irgendwie im Materiellen gegründet. Jedoch hier kann gesagt werden: mein Leib existiert nicht ohne das Bewusstsein.

Es wird jetzt klar dass in diesen Bestimmungen des Transzendenzbegriffes bei Husserl schon die ganze weitere Theorie, die man dann den "transzendentalen Idealismus" Husserls nennt, im Keim enthalten ist. Mann könnte natürlich sagen: es ist so wegen der allgemeinen Übereinstimmung. Es muss aber gezeigt werden, dass das "absolute Sein", das Nichtabsolute, das Blosstranszendente und das Blossintentionale eben so ist und nicht anders. Es muss in den Analysen der Wahrnehmung, des Bewusstseins und der Erkenntnis überhaupt gezeigt werden. Zu dieser Analyse muss man jetzt einen Weg finden, eine Methode dass man dieses reine Bewusstsein in sich selbst vor sich hat und ein Erfassen möglich wird.

Die phänomenologische Reduktion hat uns ausschliesslich auf dieses "absolute Sein" des Bewusstseins geführt.

ANMERKUNGEN

* The present paper is a part of Roman Ingarden's course given in Oslo, 1966 edited for publication by The editor of *Analecta* upon the author's permission given in 1967.

[1] Lateinisch braucht man nicht "ego cogito" zu sagen, in der grammatischen Form ist schon das "ego" da. Der Deutsche muss sagen "ich sehe", "ich erlebe", "ich denke", "ich freue mich" usw., weil der Begriff gewissermassen schon gespalten ist. Im Lateinischen ist alles ganz einfach gesagt, das "ich" ist schon dabei und es kann nicht anders sein wegen des Sinnes dieser ursprünglichen Einheit.

[2] Als ich im Jahre 1916 meine Abhandlung über Bergson schrieb, hatte ich schon nach den Studien bei Husserl und nach der Lektüre der Bücher Bergsons den Zugang zu den ursprünglichen Empfindungsdaten. Ich wusste, was sie sind und ich wusste auch, dass Husserl sie für "reelle Bestandstücke" des Bewusstseins hielt. Ich habe fast das ganze Jahr 1917 mit Husserl darüber diskutiert und immer wieder die Frage gestellt: Wie ist es mit den ursprünglichen Daten? Sind sie "reelle Bestandstücke des Bewusstseins"? Das, was mir Husserl damals zugegeben hat, war bloss dies: "Die Daten sind ich-fremd, sie sind nicht ich-lich, nicht ich-haft. Die Akte sind ich-haft, das heisst sie entspringen dem Ich, sie stammen von mir, von meinem Ich. Der Akt ist so etwas wie eine Fontäne, die aus dem Ich hervorquillt. Die Daten haben nicht die Struktur des Ich-lichen, sie sind "ich-fremd". Es ist etwas, was mich sozusagen bedrängt; eventuell ist es etwas, was nicht durchlebt wird. Die Empfindungsdaten haben aber die Struktur des ursprünglichen Bewusstseins nicht. In dem Sinne ist ein Datum kein Akt. Aber es ist doch ein reelles Stück des Bewusstseins: "das habe ich". Und so ungefähr ist es in unseren Gesprächen geblieben.

[3] Vgl. S. 60–62.

[4] Bekanntlich haben analoge Tatsachen in anderen physikalischen Experimenten Helmholtz dazu bewogen, die sinnliche Wahrnehmung, bzw. das in ihr Gegebene ganz allgemein als einen Erkenntnisvorgang aufzufassen, der dem Wahrnehmenden nur *Zeichen* von den physikalischen Tatsachen liefert und nicht die Dinge in ihren Eigenschaften selbst. Erst Husserl ist dieser Zeichentheorie der Wahrnehmung entgegengetreten.

[5] *Nota bene*, Carnap ist übrigens ein Freiburger und hat bei Husserl studiert.

[6] Das ist nach der ersten Auflage, S. 62, Zeile 2.

[7] Die letzte Wendung, die nach dem Wort "Seinsart" steht, stammt nicht aus der I. Auflage der *Ideen*, sondern ist ein Zusatz, der aus einer Randbemerkung Husserls aus seinem III. Handexemplar der *Ideen* stammt. Es ist aber durchaus im Sinne des Standpunktes Husserl in der I. Auflage der *Ideen*. Die in der Ausgabe der *Ideen* im Bd. III der Husserliana noch nachfolgenden Worte: "So verfällt es mit allem ihm Eigenen der phänomenologischen ἐποχή" ist ein weiterer Zusatz aus der soeben erwähnten Randbemerkung Husserls. Sie enthüllen nur was – und insbesondere welcher Begriff der Transzendenz – Husserl bei der Angrenzung dessen, was der "phänomenologischen ἐποχή verfällt und was als das Residuum bleibt, leitete. Das ist eben derjenige Begriff der Transzendenz, den ich jetzt im Text zu entwickeln suche.

[8] S. 69 in der ersten Auflage.

[9] Zusatz aus dem III. Handexemplar Husserls.

[10] *L.c.* S. 69.

[11] Dieser Satz steht ebenfalls nicht in der ersten Auflage der *Ideen*, sondern ist ein Zusatz, der aus einer Randbemerkung Husserls im II. Handexemplar der *Ideen* steht.

ALPHONSO LINGIS

INTENTIONALITY AND CORPOREITY

The soul's reality is based upon corporeal matter, not the latter upon the soul. (*Ideen*, III, p. 117)

By absolute consciousness Husserl did not mean to designate simply an epistemological function; absolute consciousness is a region of reality (albeit the proto-region), an ontic existent, individualizing itself across its internal temporality as a singular ego. In addition it inheres in a body.[1] This means that psychism is apperceived in the heart of Nature. But it also means that corporeity is apperceived within intentionality itself.

I. CONSCIOUSNESS NATURALIZED IN A BODY

Bodies are discovered in the course of our survey of transcendent Nature. They occupy sites in Nature's space and time – the time which is physically measured.[2] They are empirical unities of Nature, given primordially across sensory perspectives, and elaborated across the empirical-objective investigation pursued in the natural sciences. It is these things (*Körper*) that are animated by the reification or naturalization of consciousness. (Thus Husserl does not first seek the operation of consciousness in bodies by seeking to unravel the experience I have of *my own* consciousness at work in *my own* body (the *corps propre* of Sartre and Merleau-Ponty).)

The fully transcendent, empirical naturality of bodies makes them able to function as support for an emergence of psychism in Nature. Bodies appear perspectively as 'sensible schemata' connected circumstantially-causally to concomitantly varying natural things of their environments, and eventually to the whole of Nature. Along the course of these sensible apparitions, and systematically connected to them, there is apperceived the non-perspectival stream of psychic states open to empirical observation according to the methods of psychology. Husserl admits two sorts of connections between corporeal states and psychic properties: psychic properties such as sensations, affects, impulses can be functionally linked

Tymieniecka (ed.), Analecta Husserliana, Vol. I, 75–90. *All Rights Reserved Copyright* © 1970 *by D. Reidel Publishing Company, Dordrecht-Holland*

to corporeal states as to their conditions,[3] or corporeal movements may be functionally linked to psychic sequences as their expression.[4] These psychic 'properties' – functions, aptitudes, character traits, dispositions – inhere in an ego-substrate which, if not given perspectively,[5] is nonetheless manifested progressively across the course of apparition of psychic properties[6] in a unity which is not that of a sensible schema, but of a stream – a temporal rather than a spatial *Gestalt*. The psyche is thus a real (realized) ego-subject, a self-identical empirical unity manifested in individual real properties, 'interwoven' (*verflochten*) with Nature in that the course of psychic properties are systematically linked to bodily 'circumstances' (*Umstände*) by a 'functionality regulated by laws.'

Thus natural, objectal corporeity, by a sort of overdetermination of properties, supports upon itself a certain manifestation – an appresentation – of psychism in the sphere of transcendence. But the psychism 'interwoven' with bodies, 'functionally linked' to bodily circumstances, is not presented simply to the perceptual intention that primally opens upon Nature. The transgression of bodily appearances toward their 'appresented' psychic properties requires, in addition, the lateral operation of empathy (*Einfühlung*).[7] This means that the intentionality that discerns, apperceives, consciousness in Nature interwoven with bodies does so on the basis of the resemblance of those bodies *with the body with which it is itself interwoven*. There was thus first an intimate experience of a union of the corporeal with the intentional within, which was then, across the phenomenon of spontaneous corporeal association (*Paarung*) transferred (*überträgt sich*) to all analogous bodies in Nature.

The animation of bodies in Nature, observed across natural perception, thus answers to a corporeity of consciousness, to be discovered within by transcendental reflection.

No doubt it sounds strange to speak of an incarnation of consciousness in this philosophy, whose unremitting intention is to free the theory of knowledge from every kind of naturalism by regressing to the sphere of pure or absolute consciousness, where the essence of mentality is intentionality, and where intentionality is not any kind of movement of or from one real being to another (for example, from the subject to the object), but is the purely ideal 'movement' ascribing flux to identity, givenness to ideality – movement from the sensible to the intelligible. And yet corporeity appears impossible to conjure, and reappears – in a totally denaturalized and deobjectified incidence – after all the reductions, within the absolute sphere itself, as the incarnation of absolute consciousness. It is this 'absolute' corporeity that seems to us the most extraordinary of Husserl's findings regarding the rôle of the body. For a philosophy of incarnate consciousness is not genuinely formulated by asserting that the consciousness moves a body (living and operating on the outside, in the spectacle of Nature), or that a body is the vehicle of consciousness, but by showing – or discovering – by internal examination of reduced consciousness a corporeity *within*, by showing how corporeity is implicated in the internal structure of what is meant by consciousness.

What then is the intimate experience of corporeity that consciousness knows within itself?

II. THE EMERGENCE OF CORPOREITY IN THE SPHERE OF ABSOLUTE CONSCIOUSNESS

In three ways Husserl uncovers a certain corporeity – nowise objectified, nowise naturalized, nowise transcendent – implicated within the absolute sphere itself, implicated in the very operation of intentional constitution.

1. A certain corporeity of the subject is discovered in the object-structure of objects – not as a component or ingredient of objects, but as a factor of the objectivity-structure *with which* or according to which consciousness constitutes objects. Husserl uncovers in two ways such a corporeity of the subject implicated in the object-structure of objects: (a) the things are relative to a body in the polarity of their orientation, and (b) 'real' and 'unreal' properties of things are relative to a set of normal or abnormal 'states' of the subject which are an inevitable part of the total context of 'circumstances' within which sensorial appearances vary and things take form as invariants.

(a) The object-unity of things is intended across the series of perspectival sensible apparitions. The disparate succession of sensible apparitions can motivate and legitimize their coalescence into an object-unity because the intrinsic structure and grain of each is compatible with and conforms with that of the others. But in addition disparate sensible apparitions caught across different moments of time can be apparitions of different sides of one and the same thing because each is an apparition of that thing from a different point of view. Each has an index of orientation. Each aspect refers to other aspects by referring to a center of orientation that refers to a system of other centers of orientation. Thus the constitution of objects requires an intentionality that is situated somewhere – within the very space it constitutes. And the ascription of a multiplicity of sensible apparitions oriented differently to one object-identity requires the variability of the situatedness of intentionality.[8]

The situatedness and free mobility of consciousness is felt within consciousness itself in the form of kinesthetic sensations. But Husserl does not treat kinesthetic sensations as discrete representations or motor images of body movements recorded or reflected in the immobility of subjectivity. The field of kinesthetic sensations is taken as part of the total system of moving 'circumstances' in the midst of which poles of object-identity take form; they supply the orientation index which motivates the coalescence of a series of sensible appearances into one object.[9] Thus they function not as representations of objects (such as the body-object), but as motivations for the objectification of objects. Kinesthetic sensations make felt, within the sphere of consciousness, the relativity of each perceptual profile to a certain orientation. They reveal that the intentionality that means or intends an object is also a movement of *orientation* with regard to that object.

If I move my eyes, my hand, my stance, I acquire other views of the things: each profile is oriented in a context of possible movements and possible positions. The awareness of this motivation is not the performance of a reasoning; it is the kinesthetic awareness of a free variability of orientation, the free mobility of the subject, motivating the coalescence of a series of views into one object.[10] The 'if ... then ...' conditionality is the very *mode* by which kinesthetic sensations are 'kinesthetic' and not representational.[11]

Thus the crystallization of object-units out of a dispersion of differently oriented sensorial apparitions cannot fail to reveal a kinesthetic field, functioning in the constitution of every object. The field of kinesthetic sensations is not a representation, within the subject, of a materiality; it is the auto-disclosure of the materiality of the subject. It reveals subjectivity itself as oriented, already turning in the very space it will constitute by constituting objects.[12]

(b) The corporeity of the subject is also revealed in the difference between the constitution of real objects, objects really perceived, and simulacra or phantasms. For this difference involves a difference between normal and anomalous 'states' of the perceiver.

The invariant structure of a sensible thing is discerned across the course of sensorial apparitions varying in function of the context of 'circumstances' (*Umstände*). But certain variations cannot be ascribed to circumstances given in the phenomenal field, and thus, not integrated into the field of determinable phenomenal links, do not acquire solid reality, remain phantasmal.[13] When they are affected, compensated for or eliminated, by changes of position or attitude, they are rejected from the system of sensorial properties of a thing and ascribed to the material state of the subject. "With each flutter of my eyelashes a curtain lowers and rises, though I do not think for an instant of imputing this eclipse to the things themselves; with each movement of my eyes that sweep the space before me the things suffer a brief torsion, which I also ascribe to myself; and when I walk in the street with eyes fixed on the horizon of the houses, the whole of the setting near at hand quivers with each footfall on the asphalt, then settles down in its place."[14] Thus the underbrush of sensorial phantasms are denounced as such by the 'true' apparition of the things, and disclose the set of subjective adjustments that will conjure them from the world. These adjustments are not adjustments of a hypothetic *Sinngebung* being repudiated by the non-compatibility of a series of sensible appearances; they are rather material adjustments – adjustments of position, stance, or state. In these adjustments, and in the necessity for them, there arises the evidence of a state of normal corporeity of the subjectivity ("psychophysical conditionality"[15]), subject to anomalous states.

Thus the possibility of intending objects across the flux of sensorial apparitions involves the disclosure of corporeity in the form of a "position" and "state" of the subject. Consciousness is *localized* in this position and in this 'state'. The localization of consciousness, *implicated* by the panorama of the things, is revealed *within* by the states Husserl names *Empfindnisse*.[16] *Empfindnisse* are states of contact (touch, pressure, heat, cold, etc.) and of 'being moved' (in both the kinesthetic and the affective sense: states of tension and relaxation, pleasure, pain, agreeableness, disagreeableness, etc.). They are states in which the distinction between subject and object, between the feeling 'act' and the felt 'content', is not yet discernible; in them an ambivalent sensoriality reflects both the sensible qualities that announce the presence of a transcendent thing and the sensitivity of the perceiver localized in an 'extensivity'. *Empfindnisse* 'spread out' from within in an extensivity (*Ausbreitung*) of sensorial content which contrasts with the extension (*Ausdehnung*) of the transcendent objects intended across them.[17] They reveal consciousness in a voluminosity, in hands that feel the heat, feet that feel the cold, fingers that feel the relief and grain of things[18]; they reveal consciousness as a sensitivity that inhabits organs that move.

This first analysis of the meaning of corporeity in consciousness has been exterior and retrogressive; it consists in surveying the field of objectivity in which objects take form, and in discovering the subjectivity as part of the 'circumstances' which form the context of every object. Corporeity here means that every consciousness of a space of objects has itself a position within that space, and every consciousness of real things and of phantasms has a normal, and an anomalous, 'state'. The analysis of the ambivalence of *Empfindnisse* reveals an internal experience of the consciousness localized in that position and state.

2. If we turn now to the Husserlian explanation of intentionality we shall discover an incidence of corporeity revealed through direct reflection on the essence of consciousness.

Objects are the telos of consciousness; they are attained in intuition, wherein their 'carnal' (*leibhaft*) presence is immediately *given.* But for them to be given is not for objects to be received into some receptacle; consciousness orients itself toward objects, and is synthetically receptive of them, inasmuch as it posits or presumes (*meinen*) the ideal identity regulating the indefinite differentiation of their sensorial aspects. The givenness of the object is in the ideal identity that is maintained and affirmed across the flux of conscious attitudes and states in its regard.

But objects are (perhaps always[19]) intended across the materiality of hyletic data. Hyletic data are 'sensations'. They are intra-psychic, they belong to the sphere of reduced consciousness, and can be discerned by a kind of reflection.[20]

The Husserlian doctrine on sensations is expressed through a conjuncture of a hylomorphic conception of the mind – undermined – with a hermeneutic conception of the mind – undermined. On the one hand, the 'sensations', the hyletic data, are mental 'material'; they are to intentionality as matter is to form, and intentionality constitutes objects by 'animating' (*beseelen*, *durchgeistigen*) this matter. Intentionality is the formative animation of psychic matter. And yet the patterns of hyletic matter (they do form 'patterns'[21]) animated by the intentional operation of the mind, forming 'representatives' of things in the mind, in no way forms 'mental representations' of things; it is not these patterns that the mind knows, consciousness is not consciousness *of* these hyletic patterns. The objects of consciousness are not representations 'formed' out of this matter. (This is the fundamental difference between the phenomenological and the empiricist conceptions of sensations.)

On the other hand, Husserl uses a hermeneutic schema of the mind – undermined. He says that intentionality is an 'interpretation' (*Deutung*) of sensations in the mind, 'taking them to mean' the presence of transcendent objects. The 'movement' of intentionality which goes forth unto the things themselves is the purely ideal movement of interpretation of intrapsychic data. It is the purely ideal movement which presumes or posits (*meinen*) poles of ideal identity across the flux of sense data, taking the sensorial *as* adumbrations *of* something intelligible. And yet this 'interpretation' is peculiar in that the data 'interpreted' are not first given in themselves. It is because they are not themselves given that hyletic data can be 'taken as' the immediate evidence of the presence of transcendent objects. (This is what keeps the phenomenological explanation from being an intellectualist conception of perception.)

The hyletic materiality is *required* both by the Husserlian conception of the things and by the Husserlian conception of consciousness. In casting these *Abschattungen*, these shadows, on the substance of the mind, things announce their carnal (*leibhaft*) presence – as 'bodies' present to some-body. And the nature of their identity – presumed, meant, but presumed on the basis of data – requires the duality of a transcendent ideality and an intra-psychic impact.

And while both the hylomorphic and the hermeneutic conceptions of the mind are maintained, conjoined, and undermined in Husserl's conception of the operation of intentionality, neither can be conceived without the intervention of sensations, of hyletic matter. The Husserlian intentionality animates/interprets this mental materiality without making of it any kind of tableau; materiality affects, encumbers the Husserlian consciousness and makes of it not just an epistemological function, but life, *Erlebnis*, experience lived through. The hyletic substance is the *corps de l'esprit* (Valéry), a corporeity *of* intentionality.

3. Starting with the object-body perceived in the sphere of transcendence, moving to the corporeity that appears as 'position' and as 'state' within the sphere of objective 'circumstances' surrounding each object as its functional context (and felt within in the form of *Empfindnisse*), moving inward to the hyletic data that subsist in the reduced sphere of absolute consciousness, we have been finding an incidence of corporeity in each stage of the regression unto absolute consciousness. For Husserl the ultimate stage of phenomenological elucidation, in which the essence of the absolute will be scrutinized for itself, is the reflection on internal temporality.

Consciousness, as a synthetic operation that can intend ideality, identity, across a passing sensorial diversity, becomes possible because it is the auto-generation of a time field, primally produced by a time phase that retains past time phases and anticipates coming time phases. The retention and protention of time phases is the tension within the present of the 'living now' which primally makes possible the intention toward identity that synthesizes the sensorial dispersion.[22]

Husserl conceives the phase of the 'living now' as an impression, and thus the field of temporality as a mutual implication of intention and impression.[23] The 'life' or 'animation' of the primary sensory contents is their continual survival in retentions of the phase of the 'living now'. The 'living now' is the *Ur-impression*: the primal impression, the origina*tive* impression. The substance of the mind is ultimately *impressional* in nature.

What is meant by such an impression? Assuredly to think the essence of the absolute sphere in the term 'impression' is to think the substance of the mind as materiality and passivity; it is to think the absolute as Fact and not as Concept.[24] But it is no longer to think that the elementary mental event can be causally linked, as impression to impressor, as effect to cause, with physical events; Husserl thinks the primal mental event as impression *after the whole of the transcendent physical universe has been reduced*. The 'living now' which he calls an impression is not an effect of a physical process; it is that in which the totality of physical processes are constituted. It is an impression that is not impressed on any prior mental substance; it is, as it were, an impression that is its own substrate.[25] It "is the absolute beginning ... the primal source, that from which all others are continuously generated. In itself, however, it is not generated; it does not come into existence as that which is generated, but through spontaneous generation. It does not grow up (it has no seed): it is primal creation."[26] It is with itself that the impressional phase is linked, in the continuous differentiation of itself by which it originates its survival in retentions and its renewal in protentions. Thus the 'living now' is originative: *Ur-impression.* But it is impression: givenness, a "passive reception, which gathers in the novel, strange, and originary ..."[27]

III. CORPOREAL INTENTIONALITY

We have seen an inevitable corporeity emerge within the reduced sphere of 'pure' consciousness itself. But corporeity here no longer means 'extension' or 'sensoriality'; it is an utterly denaturalized and deobjectified corporeity, whose essence is conceived as 'kinesthetic sensations', 'psychophysical state', '*Empfindnisse*', 'hyletic materiality', and 'impressionality'. The concept of corporeity can no longer be treated as a subspecies of the concept 'thing'.[28]

Merleau-Ponty founded his theory of the essence of corporeity in his early work upon a transposition into corporeity of the very format of intentionality itself. He sought to show, in *The Structure of Behavior* and *Phenomenology of Perception*,[29] that intentionality, Heideggerianized as 'existence', is enacted by body movement. Intentionality will be the concept that would bring intelligibility both to bodily behavior observed physiologically and psychologically, and perception explored transcendentally. It was this, and not the elaboration of a descriptive account of a dimension of corporeity lived in the first person singular, the 'lived body' or '*corps propre*', contrasted with corporeity in the third person, the 'objective body', that was radical in his work. It is true that already for Husserl the ultimate substance of the mind is impressional in nature, and, since the whole hyletic layer involved in the constitution of objects is localized in *Empfindnisse*,[30] every intention aiming at objects is also a movement of orientation in regard to them. But in abandoning the conjugated hylomorphic-hermeneutic schemas of mental activity and in transposing the format of intentionality into the essence of corporeity, the intentional analytic no longer operates within a metaphysics where consciousness is to Nature in the relation of absolute and relative, origin and derived, but now operates in a philosophy of Nature where consciousness itself is produced by an 'overlapping' of Nature upon itself. Intentionality, which for Husserl was a transcending of the sensible, becomes a power of the sensible.

Merleau-Ponty has not fallen upon this theoretical program as a result of a simple naturalization of the transcendental. The pivot of this transposition is a new analysis of the ontological structure of the sensible thing.[31]

Repudiating the analysis of a thing in terms of facts and essence, sense-data and intelligible essence, qualia and eidos[32], Merleau-Ponty subjects the sensuous aspect of things itself to renewed scrutiny. Contrast with ideal being had led to relegating to the sensuous an existence in the here and the now, the ineffable opacity of a presence that is total or null. But the properly sensuous aspect of things is reducible neither to atomic sensations (*Phenomenology of Perception* was to show the inconsistency and, finally, the incomprehensibility, of the very notion) nor to qualia.[33] What kind of being then do the sensuous moments have? Merleau-Ponty argued that the sensuous aspects themselves – the color, the tone, the odor – are *dimensional*; they exist in trajectories of time and space and in depth.[34] In addition each has a kind of internal microstructure; red is not a blank opacity of a certain degree of density; it articulates a certain style or way of filling out, of occupying, its field – explosive, implosive, recessive, tightening, homogeneous or periodic, vibrant or absorptive, etc. Each is "a certain style, a certain manner of managing the domain of space and time over which it has competency, of pronouncing, of articulating that domain, of radiating about a wholly virtual center – in short a certain manner of being, in the active sense, a certain *Wesen*, in the sense that, says Heidegger, this word has when it is used as a verb".[35]

This concept of the inner being of the sensuous aspect permits Merleau-Ponty to conceive in a new way the *Gestaltung* that produces a thing. A thing can now be conceived as a set of sensuous aspects which each, in its own register of sensoriality, reflects the microstructure and style of all the others. One 'manner of occupying space and time' – thinly, symmetrically, periodically, tautly, etc. – reverberates across all registers of sensoriality, answers to each kind of sensorial sampling. Thus the essence of the thing is itself sensorial.

Merleau-Ponty was strongly influenced by Saussurian linguistics, and in his late work sought to express the relations between things and the relations between things and their meanings according to the model of the linguistic field analyzed structurally. The phenomenal field must then be conceived as a "diacritical, relational, oppositional system",[36] which is 'phrased' into things by systematic internal processes of articulation. Thus things do not take form about ideal poles of identity posited by the intentionality of consciousness, around which sensorial profiles would be assembled; rather things are engendered by a relief of the divergence each marks from the others – they take form by systematic lateral differentiation rather than by vertical identification. It is the lateral and oppositional relationship of one term to another that makes each of them significant, such that meaning appears at the intersection of and as it were in the interval between them.[37] And thus, just as one does not learn the significance of language by learning the meaning of words one by one, additively, so one cannot apprehend the phrasing of the phenomenal field into things by 'acts' which posit the identity of each.

How then are such beings perceived? The sensuous datum can no longer be conceived as ineffable moments of opacity reflected against the mind, or as matter formed by the mind. Each sensuous phenomenon is rather a specific articulation of a field of space and a trajectory of time, and the sensibility receives its presence within itself by actively taking up its vibrancy and internal microstructure, by actively conforming with it as upon "rays of the world",[38] by eye movements, by exploratory movements of touch, by shiftings of postural schema.[39] To see it is to see according to it, to see with it. The sensuous is both what we see, and the dimension into which we see. If the sensuous aspect itself is conceived as dimensional, its reception must be enacted in a *movement* across space and time, by a being whose own substance is able to conform to the style and grain of the things, and thus receive within itself the echo of their structures.[40] Such a being must then not be defined as the metaphysical opposite of the beings it knows (as subjectivity is defined by its op-position to ob-jectivity); it must be of the same sort as the being it knows, sensible like it.[41] "My body is to the greatest extent what every thing is: a *dimensional this*. It is the universal thing – But, while the things become dimensions only insofar as they are received in a *field*, my body is this field itself, i.e. a sensible that is dimensional *of itself*, universal measurant – "[42]

But if the sensible knower then is able to conform itself to one of the aspects of the thing, one of these 'rays of the world', it already is in the presence of the thing itself. The thing *is* nothing but a configuration in which all the sensorial aspects implicate and mutually express one another. Transcendence is displaced, in the thought of Merleau-Ponty, from being the defining characteristic of consciousness, intentionality, to the 'flesh' of the sensible as such, whose being is to be "always further on ..."[43] The sensible "is nothing else than a brief, peremptory manner of giving in one sole something, in one sole tone of being, visions past, visions to come, by whole clusters".[44] We *see*, we *touch* the thing itself; the unity of a thing is not a presumption (*Meinung*) of consciousness, but is exhibited in the fabric of the sensible itself. The structural complex which is a thing, intersensorially observable, is not reached by an 'interpretation' of sense data which would 'adumbrate' it; it is not an ideal unity of aspects which would be reached by an intentional act transcending the sensible toward the ideal. For Merleau-Ponty the passage from one sensuous aspect to the thing itself is effected by the convergence of all the receptive powers of the body upon it, a convergence effected by the synthetic, or synergic, unity of the body schema. "The whole Husserlian analysis is blocked by the framework of *acts* which imposes upon it the philosophy of *consciousness*. It is necessary to take up again and develop the *fungierende* or *latent* intentionality which is the intentionality within being. That is not compatible with 'phenomenology', that is, with an ontology that obliges whatever is not nothing to *present* itself to the *consciousness* across *Abschattungen* and as deriving from an originating donation which is an *act*, i.e. one *Erlebnis* among others ... . It is necessary to take as primary, not the consciousness and its *Ablaufsphänomen* with its distinct intentional threads, but the vortex which this *Ablaufsphänomen* schematizes, the spatializing-temporalizing vortex (which is flesh and not consciousness facing a noema."[45] As the telos of the convergence of corporeal sensitivity, and not an ideal essence comprehended in its coherence, the thing-structure remains the term of progressive exploration, retains its transcendence and its opacity.[46]

The concept of intentionality one has is commanded by the conception one has of the internal structure of the object, telos of intentionality. Merleau-Ponty exhibits the operation of intentionality as a receptive movement of a being whose metaphysical constitution is like that of the being received – sensible like the world; intentionality (he rather says "existence") is enacted by body motility, it will be the very essence of corporeity. Merleau-Ponty's first two works endeavored to show that this is not only true of the corporeity discovered within, by transcendental reflection on the reduced sphere of absolute consciousness, but also true of the corporeity observed without, in the form of behavior opening upon a 'world'.

IV. CONCLUSION: THE 'MIND'S BODY' AND NATURE

We have seen that the concept of intentionality, as exposed by Husserl and by Merleau-Ponty, requires a denaturalized, deobjectified conception of an 'absolute' corporeity – the 'mind's body' (*corps de l'esprit*). In both cases the 'mind's body' is essential to the conception of the essence of mentality; the concept of intentionality is then not a 'spiritual' or 'mental' concept, and in conceiving the essence of consciousness in the notion of intentionality we do not conceive a mental activity which would then have to be linked to a body, by compound or by composition.

It is true that this corporeity is conceived differently by the two philosophies, and this difference is commanded by their divergent understandings of what is to be meant by intentionality.

Husserl conceives the operation of intentionality, across a hylomorphic-hermeneutic schema, as a receptive intuition which synthetically posits objects by transgressing systems of data toward their ideal identity. The motility of the subject is involved in every coalescence of differently oriented *Abschattungen*. This exposition involves corporeity under the concepts of the 'position' and 'state' of the subject, 'kinesthetic sensations', '*Empfindnisse*', 'hyletic data', and 'impressions'.

For Merleau-Ponty the apprehension of things (which assuredly cannot be a process of passive reduplication of them in a mental substance) is no longer an apprehension of their ideal meanings for a mind; it is a captivation, in oneself, of their 'manner of filling space and time', their sensible essence. For Merleau-Ponty already the apprehension of any sensible phenomenon (and not only the apprehension of a sequence of differently oriented sensorial aspects) must already be a movement across time and space, because all sensible phenomena exist in spatio-temporal "rays", are dimensional, and because to apprehend them is to capture in oneself the internal microstructure that governs the deployment of each.

And to Merleau-Ponty it is unintelligible to attribute this power to a being of a different metaphysical constitution than the sensible things: only a being *like* the sensible can capture in itself their essence and micro-structure ("a mind could not be captured by its own representations; it would rebel against this insertion into the visible which is essential to the seer"[47]). But then what we have called the transcendental or 'absolute' corporeity of the mind is not to be metaphysically opposed to the corporeity (*Leibhaftigkeit*) of the things themselves; the 'mind's body' is the very inscription of Nature within.[48]

This brings Merleau-Ponty to the view that consciousness is not enacted by acts and positings, but by intentional transgressions, *Überschreiten*, which are not the initiatives of an ego, but overlappings of 'Nature' – 'Being' – upon itself. The ultimate elucidation takes the form of a reflection on the primal unity of the 'Natura naturans' or 'Being' antecedent to the segregation into mind and into phenomenal Nature, and on the internal radiation or 'overlapping' upon itself upon which 'acts' and 'attitudes', including the 'act' of intentionality, are founded.

But for Husserl's ultimate intentions, the 'corporeity of the mind' itself is to be suspended upon 'acts', hyletic data are to be shown to be constituted by the 'longitudinal' system of acts by which consciousness is primally constituted as a time-field.[49] Thus the 'absolute' corporeity discovered inevitably within the mind after all the reductions of Nature does not have the sense of an inscription of Nature within the mind: Husserl will maintain to the end the conception according to which the mind is to Nature in the relation of absolute and relative, origin and derived, unity and difference.

The Pennsylvania State University

REFERENCES

[1] "Every perceiving consciousness has this peculiarity, that it is the consciousness of *the embodied (leibhaftigen) self-presence of an individual object* ...." Edmund Husserl, *Ideas* (transl. by W. R. Boyce Gibson), New York 1931, p. 114.

[2] *Ideas*, p. 149.

[3] Edmund Husserl, *Ideen zu Einer Reinen Phänomenologie und Phänomenologischen Philosophie, Zweites Buch*, The Hague 1952, pp. 120–125, 154–157.

[4] *Ibid.*, pp. 236–247.

[5] *Ideas*, p. 150.

[6] *Ideen*, II, p. 216.

[7] *Ideen*, II, pp. 167–169; *Cartesian Meditations* (transl. by Dorion Cairns), The Hague 1965, §§ 50, 51.
[8] *Ideen*, II, pp. 56–58.
[9] *Ibid.*, pp. 38–41, 158–59.
[10] *Ibid.*, pp. 57–58.
[11] Cf. Emmanuel Lévinas, *En découvrant l'existence avec Husserl et Heidegger*, 2nd. ed., Paris 1967, p. 158.
[12] *Ibid.*
[13] *Ideen*, II, pp. 63–64.
[14] Maurice Merleau-Ponty, *The Visible and the Invisible* (transl. by A. F. Lingis), Evanston 1968, p. 7.
[15] *Ideen*, II, pp. 65–75.
[16] *Ibid.*, p. 146.
[17] *Ideas*, p. 216.
[18] *Ideen*, II, p. 145.
[19] *Ideas*, p. 227.
[20] *Ibid.*, pp. 264, 109.
[21] *Ibid.*, p. 119.
[22] It is the 'logitudinal intentionality' which makes possible the object-constitutive intentionality. Edmund Husserl, *The Phenomenology of Internal Time-Consciousness* (transl. by James S. Churchill), Bloomington and London 1964, p. 107.
[23] *Ibid.*, p. 92.
[24] "Am Anfang ist die Tat." Edmund Husserl, *Die Krisis der Europäischen Wissenschaften und die transzendentale Phänomenologie*, The Hague 1954, p. 158.
[25] "An impression ... is to be grasped as primary consciousness which has no further consciousness behind it in which we are aware of it." *The Phenomenology of Internal Time-Consciousness*, p. 117.
[26] *Ibid.*, p. 131.
[27] *Ibid.*, p. 115.
[28] "The flesh is not matter, in the sense of corpuscles of being which would add up or continue on one another to form beings. Nor is the visible (the things as well as my own body) some 'psychic' material that would be – God knows how – brought into being by the things factually existing and acting on my factual body. In general, it is not a fact or a sum of facts 'material' or 'spiritual'. Nor is it a representation for a mind: a mind could not be captured by its own representations; it would rebel against this insertion into the visible which is essential to the seer. The flesh is not matter, is not mind, is not substance .... What we are calling flesh, this interiorly worked-over mass, has no name in any philosophy." Maurice Merleau-Ponty, *The Visible and the Invisible*, pp. 139, 147.
[29] *The Structure of Behavior* (transl. by Alden L. Fisher), Boston 1963. *Phenomenology of Perception* (transl. by Colin Smith), New York 1962.
[30] Thus all sensations are kinesthetic. *Ideen*, II, p. 153.
[31] "The starting point = the critique of the usual conception of the *thing* and its *properties* → critique of the logical notion of the subject, and of logical inherence → critique of the *positive* signification (differences between significations), signification as a separation (*écart*), theory of predication – founded on this diacritical conception." *The Visible and the Invisible*, p. 224.
[32] *Ibid.*, p. 130.
[33] *Ibid.*, pp. 131–33.
[34] "Sensoriality: for example, a color, yellow; it surpasses itself of itself: as soon as it

becomes the color of the illumination, the dominant color of the field, it ceases to be such or such a color, it has therefore of itself an ontological function, it becomes apt to represent all things (like engravings, *Dioptrics*, Discourse IV). With one sole movement it imposes itself as particular and ceases to be visible as particular." *Ibid.*, pp. 217–18.

[35] *Ibid.*, pp. 115.

[36] *Ibid.*, p. 213.

[37] Maurice Merleau-Ponty, *Signs* (transl. by Richard C. McCleary), Evanston 1964, p. 42.

[38] *The Visible and the Invisible*, pp. 241–42, 218.

[39] *Phenomenology of Perception*, pp. 208–212; *The Visible and the Invisible*, pp. 134–135.

[40] "Between the exploration and what it will teach me, between my movements and what I touch, there must exist some relationship by principle, some kinship, according to which they are not only, like the pseudopods of the amoeba, vague and ephemeral deformations of the corporeal space, but the initiation to and the opening upon a tactile world. This can happen only if my hand, while it is felt from within, is also accessible from without, itself tangible, for my other hand, for example, if it takes its place among the things it touches, is in a sense one of them, opens finally upon a tangible being of which it is also a part." *The Visible and the Invisible*, p. 133.

[41] "It is that the thickness of flesh between the seer and the thing is constitutive for the thing of its visibility as for the seer of his corporeity; it is not an obstacle between them, it is their means of communication." *Ibid.*, p. 135.

[42] *Ibid.*, pp. 259–260.

[43] *Ibid.*, p. 217.

[44] *Ibid.*, p. 135.

[45] *Ibid.*, p. 244.

[46] *Phenomenology of Perception*, pp. 322–327.

[47] *The Visible and the Invisible*, p. 139.

[48] "The meaning of being is to be disclosed: it is a question of showing that the ontic, the '*Erlebnisse*', 'sensations', 'judgments'– (the objects, the 'represented', in short all idealizations of the Psyche and of Nature) all the bric-à-brac of those *positive* psychic so-called 'realities' (and which are lacunar, 'insular', without *Weltlichkeit* of their own) is in reality abstractly carved out from the ontological tissue, from the 'mind's body' -- Being is the 'place' where the 'modes of consciousness' are inscribed as structurations of Being ..., and where the structurations of Being are modes of consciousness." *Ibid.*, p. 253.

[49] *Ideas*, p. 226.

ULRICH CLAESGES

INTENTIONALITÄT UND TRANSZENDENZ

*Zur Konstitution der materiellen Natur**

Die Idee der Intentionalität entwickelt ihre Fruchtbarkeit erst in Verbindung mit dem methodischen Grundprinzip der Phänomenologie Husserls: der phänomenologischen Reduktion. Dieses Prinzip entwickelt Husserl 1907 im Zusammenhang mit dem erkenntnistheoretischen Problem der Transzendenz. Im folgenden soll untersucht werden, was die Idee der Intentionalität im Hinblick auf das Problem der Transzendenz zu leisten vermag. Wir wollen uns dabei jedoch auf einen speziellen Problembereich beschränken, nämlich auf die Konstitution der materiellen Natur.

Das erkenntnistheoretische Problem der Transzendenz wird keineswegs erst durch die Phänomenologie aufgeworfen; vielmehr findet sie es vor im Selbstverständnis der natürlichen Erkenntnis und in einer Fülle daran sich anschließender Lösungsversuche. Dabei entspringt das Problem der Transzendenz einer Reflexion auf die Unterscheidung von wirklicher und bloß vermeintlicher Erkenntnis. Beide sind, als Erlebnis des erkennenden Subjekts genommen, in gleicher Weise real und demnach als solche nicht zu unterscheiden. Ihr Unterschied besteht vielmehr darin, daß der wirklichen Erkenntnis ein Gegenstand, das Erkannte, entspricht; dieser ist nicht dem Erkenntniserlebnis immanent, sondern ihm gegenüber transzendent. Der bloß vermeintlichen Erkenntnis dagegen entspricht kein solcher ihr transzendenter Gegenstand. Von daher könnte das erkenntnistheoretische Problem der Transzendenz so formuliert werden: Wie kann ein dem Subjekt immanentes Erlebnis einen in diesem Erlebnis nicht enthaltenen Gegenstand treffen?

Diese Frage stellt das natürliche Bewußtsein gewöhnlich nicht. Ihm ist die transzendent-objektivierende Leistung der Erkenntnis selbstverständlich. In dem Augenblick aber, wo das natürliche Bewußtsein diese Frage aufwirft, verwickelt es sich, wie Husserl gezeigt hat, in Widersprüche. Sie haben ihren Grund darin, daß dasjenige, was zum Problem gemacht wird, nämlich die transzendente Leistung der Erkenntnis, auf der anderen Seite als selbstverständlich in Geltung bleibt. Genau diesen Fehler vermeidet die Phänomenologie. Sie nimmt das Problem der Transzendenz

Tymieniecka (ed.), Analecta Husserliana, Vol. I, 91–99. *All Rights Reserved*
Copyright © 1970 *by D. Reidel Publishing Company, Dordrecht-Holland*

in der Weise auf, daß sie die genannte, im Selbstverständnis der natürlichen Erkenntnis implizierte Selbstverständlichkeit ausschaltet. Dies geschieht in der phänomenologischen Reduktion. Damit ist die Transzendenz nicht beseitigt, sondern sie bleibt als Anspruch der natürlichen Erkenntnis, der nun geprüft werden muß, erhalten. Die Prüfung dieses Anspruches setzt an bei einer Analyse der intentionalen Struktur des Bewußtseins, wobei sich zeigt, daß das natürliche Erkennen das Problem der Transzendenz gar nicht zureichend gestellt hat. Es muß nämlich in einer doppelten Weise zwischen Immanenz und Transzendenz unterschieden werden.

Ein erster Unterschied zwischen Immanenz und Transzendenz kommt jedem Bewußtsein zu; es ist der zwischen den Akten des Vermeinens und dem Vermeinten als solchem. Das Vermeinte ist in den Akten des Vermeinens nicht reell enthalten, sondern ihnen gegenüber transzendent. Dieser Unterschied zwischen cogitatio und cogitatum, zwischen noesis und noema ist als Wesensbestimmung der Intentionalität bekannt; wir brauchen ihn hier nicht zu erläutern.

Der zweite Unterschied von Immanenz und Transzendenz betrifft das Reich des Noematischen selbst. Hinsichtlich eines bestimmten noema kann nämlich völlige Identität des Gegebenen und Gemeinten bestehen: Das Gemeinte ist genau so gemeint, wie es gegeben ist. Diese Identität des Gegebenen und Gemeinten faßt Husserl als Selbstgegebenheit, eine Gegebenheit, die jeden Zweifel ausschließt. Husserl nennt sie Perzeption.

Meist geht aber in einem noema das Gemeinte über das hinaus, was von ihm selbstgegeben ist. Diesen Überschuß der Meinung über das jeweils Selbstgegebene nennt Husserl Apperzeption. Das in der Weise der Apperzeption Gemeinte ist gegenüber dem in der Perzeption Selbstgegebenen in einem neuen Sinn transzendent. Die Unterscheidung von Immanenz und Transzendenz besagt hier also die Unterscheidung von Perzeption und Apperzeption.

Mit dieser Unterscheidung im Bereich des Noematischen selbst ist diejenige Wesensbestimmung von Intentionalität gewonnen, die es gestattet, das Problem der Transzendenz zureichend zu stellen. Da der Unterschied von noesis und noema, und damit die Transzendenz des Noematischen als solchen, jedem Bewußtsein zukommt, muß das erkenntnistheoretische Problem der Transzendenz als das Problem des Verhältnisses von Perzeption und Apperzeption aufgefaßt werden.

Allgemein kann zum Verhältnis dieser beiden soviel gesagt werden, daß sie in einem notwendig geregelten Zusammenhang stehen, und zwar so, daß eine Apperzeption nur auf dem Boden einer Perzeption möglich ist. Was dies im einzelnen besagt, wird deutlich, wenn wir nun das Problem der Apperzeption im Hinblick auf die Konstitution der materiellen Natur erörtern. Dazu ist jedoch eine Vorbemerkung notwendig.

Die Konstitution einer Region von Gegenständlichkeiten wird aufgeklärt in einer sogenannten Intentionalanalyse. Diese fragt von einem intentionalen Leitfaden, der in einer regionalen Ontologie erstellt wird, auf die mannigfaltigen wirklichen oder möglichen Bewußtseinsweisen zurück, in denen der als Leitfaden fungierende Gegenstand gegeben wird. Den Leitfaden für die Konstitutionanalyse der materiellen Natur bietet die regionale Ontologie des Dinges. Wie bekannt, unterscheidet Husserl im Wesen des Dinges drei Schichten, die in einem bestimmten Fundierungszusammenhang stehen. Die drei Schichten sind *res temporalis*, *res extensa* und *res materialis*, an deren wesentliche Bestimmungen hier kurz erinnert sei. Ein Ding ist *res temporalis*, sofern es notwendig als Einheit der Dauer in einem zeitlichen Abfluß bestimmt ist. Ein Ding ist *res extensa*, sofern es notwendig als sinnlich qualifizierte Gestalt in einer Lage bestimmt ist und unter Erhaltung seiner Identität möglichen Gestalt- und Lageveränderungen unterworfen sein kann. Ein Ding ist *res materialis*, d.h. substantial-kausales Ding, sofern es auf andere Dinge Wirkungen ausüben und von anderen Dingen Wirkungen erleiden kann bzw. sofern es Träger von wirkenden und erwirkten Eigenschaften ist.

Diese drei Schichten stehen in einem einseitigen Fundierungsverhältnis. Jede *res extensa* ist notwendig *res temporalis*, nicht aber umgekehrt; und jede *res materialis* ist notwendig *res extensa*, nicht aber umgekehrt. Die Intentionalanalyse hinsichtlich dieser drei Schichten des Dinges kann hier nicht im vollen Umfang verfolgt werden. Bezüglich der Schicht der res *extensa* habe ich dies an anderer Stelle versucht[1]. Hier wollen wir vielmehr nur das Verhältnis von Perzeption und Apperzeption in den genannten Schichten untersuchen, um so einer Lösung der Transzendenz im Hinblick auf die materielle Natur näher zu kommen.

Schon bei der Konstitution der *res temporalis* (als Beispiel diene etwa eine Melodie) haben wir ein gewisses Verhältnis von Perzeption und Apperzeption. Wir sind im Hören auf die Melodie als ganze gerichtet. In der Weise der Perzeption gegeben ist jedoch jeweils nur die urimpressionale, im aktuellen Jetzt gegebene Phase. Zur Melodie gehören aber notwendig auch schon die vergangenen Phasen, die das Bewußtsein in der Weise der Retention festhält. Die Gegebenheitsweise der so festgehaltenen Phasen kann man als Apperzeption bezeichnen; sie sind notwendig mitgemeint, obwohl nicht in der urimpressionalen Gegenwart selbstgegeben. Ähnlich verhält es sich mit den künftigen Phasen der Melodie. Sie sind in der Weise der Protention im Bewußtsein der ganzen Melodie mitgemeint, aber nicht selbstgegeben.

Das Verhältnis von Perzeption und Apperzeption unterliegt notwendig einer bestimmten Regelung. Diese Regelung ist im Fall der *res temporalis* in der Struktur des inneren Zeitbewußtseins gegeben; sie besteht darin, daß zu jedem urimpressionalen Jetzt notwendig Retention und Protention gehören. Diese Struktur regelt aber auch die Überführbarkeit von Apperzeption in Perzeption. In der ausdrücklichen, d.h. aktiven Wiedererinnerung wird die vergangene Phase, allerdings bloß als vergangene, zur Selbstgegebenheit gebracht. Anders steht es mit der in der Protention apperzipierten künftigen Phase der Melodie. Diese wird durch den passiven Abfluß der Jetztfolge selbst in Perzeption überführt. Die mitgemeinte künftige Phase der Melodie rückt von selbst in die urimpressionale Gegenwart ein.

Diese Andeutungen mögen genügen, um die Regelstruktur des Verhältnisses von Perzeption und Apperzeption sichtbar werden zu lassen: Dieselbe Struktur (hier die der immanenten Zeit), die das Verhältnis von Perzeption und Apperzeption bzw. deren notwendige Einheit regelt, regelt auch die Überführbarkeit von Apperzeption in Perzeption. Von einer *res temporalis* kann also nur das im strengen Sinn immanent genannt werden, was von ihr in der jeweiligen urimpressionalen Gegenwart gegeben ist. Vergangene und zukünftige Dauer sind demgegenüber transzendent. Dies überträgt sich wegen des angegebenen Fundierungsverhältnisses auf die anderen Schichten des Dinges. Doch soll der Einfachheit halber im folgenden davon abgesehen werden.

Die genannte Gesetzlichkeit, daß die Regelstruktur, die das Verhältnis von Perzeption und Apperzeption regelt, zugleich auch die Überführbarkeit der Apperzeption in die Perzeption, also die Überführbarkeit des bloß Gemeinten in ein Selbstgegebenes sicherstellt, wird bei der Konstitution der *res extensa* noch deutlicher. Wir gehen aus von dem Beispiel der Wahrnehmung einer Kugel. In der Weise der Perzeption ist gegeben eine Seite der Kugel, und zwar aufgrund einer bestimmten kinästhetischen

Situation, oder, naiv gesprochen, aufgrund einer bestimmten Stellung des Betrachters zum Objekt. Die Kugel ist aber gemeint als Kugel, d.h. als eine gleichmäßig allseitig geschlossene Oberflächengestalt. Das aber bedeutet, daß die nicht gegebenen Seiten der Kugel notwendig mitgemeint sind in der Weise der Apperzeption. Es ist klar, wie diese Apperzeption in Perzeption zu überführen ist: Ich brauche nur, naiv gesprochen, die Kugel von allen Seiten anzusehen.

Was ist nun hier die Regelstruktur, die das Verhältnis von Perzeption und Apperzeption bestimmt? Sie ist das kinästhetische Gesamtsystem als ein System subjektiver Vermöglichkeiten. Die aktuelle kinästhetische Situation ist ja nur die Aktualisierung einer dieser Vermöglichkeiten. Dabei muß das kinästhetische Gesamtsystem als ein ideelles System angesehen werden: es hat den Charakter der Allzeitlichkeit, da einzelne Situationen in ihm jederzeit und beliebig oft realisierbar sind. Da dem kinästhetischen Gesamtsystem als Korrelat notwendig das zugehört, was wir Raum nennen, kann man auch sagen: Die Struktur des Raumes selbst regelt das Verhältnis von Perzeption und Apperzeption, und zwar in der angegebenen doppelten Weise. Diese Regelung besagt, daß das Bewußtsein von Räumlichem seiner Struktur nach notwendig unaufhebbare Einheit von Perzeption und Apperzeption ist. Zwar besteht die ideelle Möglichkeit, jede Apperzeption auf eine a priori geregelte Weise in Perzeption zu überführen; das bedeutet aber nicht, daß die Verbindung von Perzeption und Apperzeption überhaupt aufhebbar wäre. Aller Übergang von Apperzeption in Perzeption ist zugleich ein Übergang von Perzeption in Apperzeption. Diese Gesetzlichkeit begründet die Transzendenz der *res extensa*.

Wenden wir uns nun noch der Schicht der *res materialis* zu, mit der sich erst die Konstitution der materiellen Natur vollendet. Wir gehen wieder von unserem Beispiel der Wahrnehmung einer Kugel aus. Ein solches Wahrnehmungsding ist nicht bloß als *res extensa*, oder – wie Husserls Name für die qualifizierte Raumgestalt lautet – als Phantom gemeint, sondern als substantial-kausales Ding im angegebenen Sinn. Um das Verhältnis von Perzeption und Apperzeption in der Schicht der Materialität zu bestimmen, müssen wir auf das Verhältnis von *res extensa* und *res materialis* achten. (Ich folge hier den §§ 15–17 der *Ideen* II [2], ohne wegen der Kürze der Zeit, wie auch bisher schon, die Belegstellen im einzelnen zu nennen.) Wir hatten gesagt, daß jede *res materialis* notwendig *res extensa* sei, nicht aber umgekehrt. Das bedeutet, daß Phantome prinzipiell anschaulich gebbar sind, die nicht als materielle Dinge gemeint sind bzw. gemeint sein können. Eine Abstraktion in diesem Sinn lag unserer bisherigen Betrachtung zugrunde. Wir können diese Abstraktion aber noch von einer anderen Seite ansehen.

Wir nehmen unser Ding genau so, wie es zunächst aufgefaßt wird, nämlich als materielles Ding. Schalten wir nun alle Bestimmungen aus, die als eigentlich oder ausdrücklich materielle gemeint sind, dann zeigt sich, daß sich hinsichtlich dessen, was von dem Ding selbstgegeben ist, gar nichts ändert. Das Ding steht uns als dieselbe qualifizierte, von einer bestimmten Seite gegebene Raumgestalt vor Augen. Wir können um sie herumgehen, das zunächst Apperzipierte zur Perzeption bringen etc. Damit ist aber eine entscheidende Einsicht gewonnen: Ändert sich hinsichtlich des Gegebenen nichts, wenn die Schicht der Materialität weggestrichen wird, so besagt dies offenbar: Die Schicht der Materialität hat keine ihr spezifisch zugehörige Weise von Perzeption. Das, was dieser Schicht angehört, kann nicht perzeptiv selbstgegeben sein, wie etwa die Seite eines Phantoms. Anders gesprochen: Im Übergang von der *res extensa* zur *res materialis* wird keine neue perzeptive Schicht gewonnen, wie dies beim Übergang von der *res temporalis* zur *res extensa* durchaus der Fall ist. Die Schicht der Materialität ist reine Apperzeption. Was aber bedeutet das?

Wir hatten gesagt: Keine Apperzeption ist ohne Perzeption denkbar. Das gilt auch hier. Im Fall der *res materialis* nämlich übernimmt die gesamte Schicht der *res extensa* die Funktion der Perzeption. Das Ding ist als *res materialis* gemeint, obwohl es nur als *res extensa*, d.h. als Phantom gegeben ist. Ist die Schicht der *res materialis* rein apperzeptiver Natur, besteht ferner das Wesen der Transzendenz in solcher Apperzeption, so kommt der *res materialis*, die ja das eigentliche Wesen der materiellen Natur ausmacht, Transzendenz in einem ausgezeichneten Sinn zu. Bei der Untersuchung der *res temporalis* und der *res extensa* hatte sich gezeigt, daß es eine Regelstruktur gibt, die das Verhältnis von Perzeption und Apperzeption bestimmt. Es war dieselbe, die auch die Überführbarkeit von Apperzeption in Perzeption gestattet. Ist aber die Schicht der res materialis rein apperzeptiver Natur, so heißt das: Die Apperzeption ist prinzipiell nicht in Perzeption zu überführen. Eine Regelstruktur im bisherigen Sinn kann es hier also nicht geben.

Damit hat sich unsere Frage verschärft: Was regelt das für die Schicht der *res materialis* konstitutive Verhältnis von Perzeption und Apperzeption? Die Antwort, die Husserl auf diese Frage gibt, ist bekannt. Die Materialität des Dinges weist sich aus durch dessen Abhängigkeit von Umständen. Wenn wir ein isoliertes Ding betrachten, haben wir keine Möglichkeit zu erfahren, ob das Ding wirklich ein materielles Ding ist. Diese Möglichkeit besteht erst, wenn wir das Ding in seinem Zusammenhang mit anderen Dingen betrachten. Was aber in einem solchen Zusammenhang sich verändernder Dinge zu wirklicher oder möglicher Perzeption kommt, gehört wieder nur der Phantomsphäre an. Was in ihr zu möglicher oder wirklicher Perzeption kommt, läßt sich etwa so beschreiben: Eine bestimmte Bewegung als Ortsveränderung oder Gestaltveränderung oder aber eine bestimmte qualitative Veränderung des Phantoms A ist zusammen gegeben mit entsprechenden Veränderungen des Phantoms B. Ich kann auch eventuell feststellen: Immer wenn A sich so verändert, verändert sich B so. Aufgrund dieser Gegebenheiten, so behauptet Husserl, werden A und B als materielle Dinge, d.h. als substantial-kausale Einheiten aufgefaßt. Das aber bedeutet, verkürzt gesprochen, daß alle Vorkommnisse der Phantomsphäre ihren letzten Grund in der Wechselwirkung prinzipiell unanschaulicher substantialer Einheiten haben. Das, was innerhalb der Phantomsphäre zur Selbstgegebenheit kommen kann, wird aufgefaßt im Sinn substantialer oder kausaler Eigenschaften, die ihrerseits wiederum nur Strahlen – wie Husserl sagt – eines einheitlichen Seins sind.

Daß eine solche materialisierende Apperzeption, die in Wahrheit komplizierter ist, als wir es hier darlegen können, stattfindet und immer schon stattgefunden hat, mag unbezweifelt sein. Unbezweifelt mag ferner sein, daß diese Auffassung nur dort statt hat, wo geregelte Abhängigkeiten innerhalb der Phantomsphäre bestehen. Die Frage ist aber nicht, ob diese Apperzeption statthat, sondern warum sie statthat bzw. wie sie möglich ist. Diese Frage aber hat Husserl in den *Ideen* II nicht beantwortet. Er beschreibt die Apperzeption, ohne sie hinsichtlich ihrer Möglichkeit und Notwendigkeit aufzuklären. Dies aber wäre notwendig, um dem skeptischen Einwand zu entgehen, Kausalität und Substantialität seien auf purer Gewohnheit beruhende Täuschungen.

Fragen wir abschließend nach den Gründen dafür, warum Husserl eine solche Begründung der materialisierenden Apperzeption nicht gegeben hat oder nicht geben konnte. Einer dieser Gründe liegt m.E. in Husserls Begriff der ästhetischen oder sinnlichen Synthesis. Diese ist, wie Husserl ausdrücklich sagt (*Ideen* II, 20, Anm.), auch für die Konstitution der *res materialis* verantwortlich. Was ist das Wesen der ästhetischen Synthesis? Sie wird zunächst negativ bestimmt durch ihren Gegensatz zur sogenannten kategorialen Synthesis, die ihrem Wesen nach ein rein spontaner Akt ist. Eine solche Synthesis liegt z.B. in jedem Urteil vor. Dagegen ist die ästhetische Synthesis ihrem Wesen nach rezeptiv. Sie verknüpft Elemente, die entweder selber aus einer ästhetischen Synthesis stammen, oder aber letzte sinnliche Merkmale oder Gegebenheiten sind. Entscheidend aber ist, daß die ästhetische Synthesis als eine Synthesis des Gleichartigen angesehen werden muß; so werden in der Synthesis der *res extensa*, der vollen qualifizierten Raumgestalt Aspekte, Seiten etc. zur Einheit gebracht, die alle qualifiziert-extentionale Gegebenheiten sind. Und das Ganze als Resultat der Synthesis ist ebenfalls eine qualifiziert-extentionale Gegebenheit. In dieser Synthesis gehen Perzeption und Apperzeption so zusammen, daß das Apperzipierte, sofern es in Einheit mit dem Perzipierten gesetz wird, von derselben Art ist wie das Perzipierte, d.h. es ist *a priori* als Qualifiziert-Extentionales gesetzt. Und deshalb ist die Apperzeption in Perzeption zu überführen.

Nun soll aber auch die Synthesis, die die *res materialis* konstituiert, ästhetische Synthesis sein. Das substantial-kausale Ding als Einheit ist, so behauptet Husserl, Resultat eben dieser Synthesis. Das aber ist unmöglich. Was der Synthesis, die die substantial-kausale Einheit konstituiert, als Material gleichsam vorgegeben ist, gehört der Phantomsphäre an, ist also qualifiziert-extentional. Was Resultat der Synthesis ist, nämlich das reale Ding als substantial-kausale Einheit gehört nicht der Phantomsphäre an, sondern einer völlig neuen, durch diese Synthesis allererst gesetzten Sphäre. Die Synthesis ist hier also nicht Synthesis des Gleichartigen in dem Sinn, daß die Einheit dieselbe Struktur hat wie das, was durch die Synthesis zur Einheit gebracht wird. Deshalb gibt es hier keine Überführbarkeit der Apperzeption in Perzeption. Die Synthesis der *res materialis* muß also eine andere Struktur haben. Diese hat Husserl nicht angegeben.

Aber bei Kant finden wir einen Hinweis. Er unterscheidet zwischen mathematischer Synthesis und dynamischer Synthesis. Die mathematische Synthesis ist Synthesis des Gleichartigen. Sie konstituiert Qualität und

Quantität des in Raum und Zeit erscheinenden Mannigfaltigen. Husserlsch gesprochen: die *res extensa* als qualifizierte Gestalt in einer Lage. Die Konstitution des substantial-kausalen Dinges vollzieht sich bei Kant aber in der dynamischen Synthe-sis. Substanz, Kausalität und Wechselwirkung sind dynamische Kategorien. Dabei soll natürlich nicht übersehen sein, daß bei Kant auch Quantität und Qualität Kategorien sind und als solche dem reinen Verstand zugehören.

Ohne in eine Diskussion zwischen Husserl und Kant einzutreten, scheint mir doch die Frage berechtigt, ob Husserl nicht im Hinblick auf die Konstitution der materiellen Natur eine Differenz übersieht, die Kant mit seiner Unterscheidung zwischen mathematischer und dynamischer Synthesis anvisiert bzw. beachtet hat. Jedenfalls bleibt das Verhältnis von Perzeption und Apperzeption in der Konstitution der *res materialis* dunkel; dunkel bleibt damit auch der Sinn ihrer Transzendenz. Ob die Phänomenologie Husserls andere Mittel bereitstellt, dieses Problem zu lösen, muß hier offen bleiben.

Köln

ANMERKUNGEN

* Die vorliegende Arbeit ist das deutsche Original, dessen englische Übersetzung auf dem Internationalen Phänomenologie-Kongreß in Waterloo, Canada, April 1969 vorgetragen wurde.

1 Vgl. V. Claesges, *Edmund Husserls Theorie der Raumkonstitution*, Den Haag, 1964.

2 Vgl. E. Husserl, *Ideen zu einer reinen Phänomenologie und phänomenologischen Philosophie. Zweites Buch* – Husserliana, Band IV, Martinus Nijhoff, Den Haag, 1952.

J. N. MOHANTY

HUSSERL'S CONCEPT OF INTENTIONALITY

It may not be an overstatement to say that the development of Husserl's thought can be adequately grasped if only we followed the development of his concept of intentionality. The same could also be said of the development of phenomenology after Husserl. There are no doubt some excellent studies on Husserl's concept of intentionality[1], but what is still lacking is a comprehensive account which would take into account the total *Husserliana* which are only now coming to be made available to scholars. Furthermore, we also do not have a study which undertakes to assess Husserl's concept of intentionality from the point of view of the criticisms and challenges emerging out of the writings of the post-Husserlian phenomenologists. This paper will make an attempt to fulfil these two needs but only within the brief compass permissible here. There will be first an attempt to explore the philosophical relationship of Husserl's concept of intentionality to Brentano's. An exposition will next be given of Husserl's concept through the various phases of its development. This will be followed by a reference to the criticisms and challenges made by the later phenomenologists like Sartre and Merleau-Ponty, and the concluding part of the paper will seek to evaluate the points made by the critics and in that connection take a fresh look at Husserl's thought against the new perspective gained thereby.

I

While making out his case for regarding intentionality as a distinguishing feature of what he calls 'mental phenomena', Brentano makes use of such locutions as 'intentional inexistence of the object', 'relatedness to a content', 'direction towards an object' and 'immanent objectivity'. However he is also aware that these expressions are not free from equivocations. The scholastic notion of 'inexistence' dominates his way of speaking about intentionality, and he even speaks of a mental state as 'containing' something as its object, or of the psychic '*Einwohnung*'. For a proper

Tymieniecka (ed.), Analecta Husserliana, Vol. I, 100–132. *All Rights Reserved Copyright* © 1970 *by D. Reidel Publishing Company, Dordrecht-Holland*

understanding of Brentano and also for a correct appreciation of Husserl's relationship to the Brentano thesis it is necessary to be clear about Brentano's uses of 'object' and 'content' and to ask: (a) did Brentano, like Twardowski and Meinong, distinguish between object and content, and also (b) did he regard intentionality as a relation? With regard to (a), it would seem that Brentano did at first draw some kind of distinction between the mere object ('*Gegenstand schlechtweg*') and the intentional or immanent object, in which case the latter (e.g. this-tree-as-an-object-of-my-present-perception) will seem to be the content having a sort of mental or quasi-mental existence.[2] However, we learn from Kraus as also from Brentano's later writings, especially from his letters, that his view on this matter underwent a radical change. In his letter to Marty, dated 1905, he not only rejects this view but strangely enough contends that he never held it. The immanent object is now said to be the same as the thing itself. When, for example, I think about a horse the object of my thought is the horse itself, not a contemplated horse. Brentano however continues to call this thing, the horse in our example, an 'immanent' object, and to hold the view that such a thing need not exist. The latter view is explained by the fact that he uses 'thing' or 'real' in a specific sense according to which a thing or a real need not exist or need not be actual (*wirkliches*). The word 'immanent' does not any longer mean 'mental' or 'quasi-mental', and is in fact a redundant qualification of 'object'; and being an object "is merely the linguistic correlate of the person experiencing having it as object".[3] We need not recapitulate here Brentano's interesting attempts – anticipating Russell's theory of incomplete symbols – to get rid of the curious *Zwischenentitäten* by distinguishing between 'autosemantic' and 'synsemantic' expressions; but one has to recognise that although he continues to use misleading language he does come to reject the so-called contents, and that the more successfully he does so, the more the approximates towards an adequate formulation of the concept of intentionality.

As regards (b), again, Brentano no doubt *speaks* of intentional relatedness. But here also he shows keen awareness of the difficulties likely to arise out of treating intentionality as a relation. Brentano tells us[4] that whereas for a genuine relation to hold good between two terms both these terms should exist, this is not the case with mental directedness. In other words, my thought may very well intentionally refer to something nonexistent (e.g. a golden mountain). Brentano therefore suggests that what we may have here is not something *relatives* but *relativliches*.[5]

It would seem then that on Brentano's view all mental phenomena but no non-mental phenomena are characterised by a peculiar, not further analysable, directedness towards an object, a directedness which is not mediated by a mental content as distinguished from the object intended, and which seems to be a relation but is not really one. In view of this, we may now turn to Husserl's criticisms of Brentano.

We may for the present leave out of consideration Husserl's comments in the *Logische Untersuchungen*[6] on the inadequacies and equivocations of Brentano's terminology. The really pertinent challenge to the Brentano thesis, however, concerns its validity as a thesis about the alleged criterion of the 'mental'. Husserl asks[7], is everything mental intentional? Or, even, are all *Erlebnisse* intentional? Are not sensations – which undoubtedly enter into intentional acts as their components – themselves non-intentional? Further, are there not objectless feelings, e.g. anxieties? Now it would seem that although in the *Logische Untersuchungen* Husserl did hold that not all that is mental is intentional, only the so-called acts being so, he gave up this reservation in the later writings. As he extended the sphere of intentionality beyond the narrow region of what he called acts, as he undertook to lay bare the intentional structure even of the sensible data in the lectures on time-consciousness, he surely came nearer to accepting the Brentano thesis that everything mental is intentional. But at the same time he seems to have moved away from the other half of the Brentano thesis and as we shall see later on, come to hold that body, and not merely the mind, is characterised by intentionality. Thus even with Husserl, as with Merleau-Ponty, that aspect of the Brentano thesis in which it offers a criterion of the mental comes to be abandoned in the long run.

The rest of Husserl's comments are consequences (i) of his total rejection of the naturalistic-causal framework for a philosophy of mind and (ii) of the fundamental methodological notion of phenomenological reduction. Brentano was operating within the framework of a naturalistic psychology which accords a pre-eminent place to the causal mode of explanation. For Husserl, the concept of intentionality demands a complete abandonment of the causal attitude in connection with whatever is intentional.[8] This change also makes way for the ideas of intentional implication and intentional unity amongst mental states and later on for the very important phenomenological notion of motivation which both takes care of and goes beyond the old empiricist notion of association. Conscious states imply each other, lead to one another, are synthesised with each other, and thus constitute a unity – both at the formal level of time-consciousness and at the contentual level – not by mechanical association, nor by logical entailment but by motivating, anticipating and fulfilling each other. To *understand* a conscious state in this sense would require following up all its intentional implications, unravelling its motivations, etc. But it is the reduction which for the first time enables us to appreciate the full significance of the notion of intentionality.[9] Brentano was of course aware that the object intended may or may not exist, and he also saw that the intended reference qua reference is not affected by the existence or non-existence of the object of reference. But this indifference cannot be fully appreciated so long as we do not give effect to the reduction by bracketing all positing of transcendent things, and do not view the object qua object i.e. as it is being intended. In other words, it is the reduction which first makes possible the key phenomenological concept of *noema*, a concept which cuts across ontological regions of individual facts and essences, the existent and the non-existent, simply because it itself is not an ontological but a genuinely phenomenological concept.[10]

Gurwitsch has shown the affinity between the method of reduction and the concept of *noema* with the gestaltist's rejection of the constancy hypothesis and the consequent notion of *theme*.[11] While thus Husserl's thesis receives confirmation from the side of the psychologist, no less interesting confirmation announces itself from the side of the linguistically oriented philosopher. It has lately been emphasised that intentional discourse – as contrasted with the non-intentional discourse – is characterised either by (i) existence independence (in the case of simple, declarative sentences with a substantival expression functioning as the object of the verb), or by (ii) truth-value indifference (in the case of non-compound sentences containing a propositional clause), or by (iii) referential opacity (in the case of cognitive verbs like 'know', 'see', 'perceive' etc.) or, finally, by (iv) a certain peculiar pattern of implication obtaining amongst the quantified forms of sentences with intentional prefixes.[12] Now it would seem clear that at least the first three criteria are closely connected with the phenomenological distinctions between the intended object and the object *qua* intended, and between the real and the intended objects. What an intentional act does entail – not logically, to be sure – is an intended object but not a real object *qua* real. It may even be that the reality of the intended object is precisely the object of belief etc., but even in such cases one has to distinguish between its reality as such and its reality as intended i.e. as believed, imagined, thought of, etc. The concept of intentionality implies neither realism nor idealism regarded as ontological theses: it is this ontological neutrality whose linguistic correlate is brought out in the first two criteria, namely in the alleged facts of existence independence and truth value indifference.

However, while thus Husserl did in many ways go beyond Brentano and refined and improved upon his concept, many of the criticisms made by phenomenologists against Brentano are far from being fair. It has for example been contended[13] that Brentano knows only of intentional relation between two entities, and – as a corollary of this – that Brentano's concept of intentionality is based on an assumption of epistemological realism. Now it is surely true that Brentano was a realist but it has also to be admitted that in the formation of his concept of intentionality, especially in its more mature form, he did not let this belief influence him. He does say, for example, that the existence of the object is not necessarily entailed by the fact that it is the object of a certain mental directedness. He did realise the difficulties that arise in case one understands intentionality as a relation. Phenomenologists would do well to take Brentano's later theories and self-criticisms more seriously.

Again, Herbert Spiegelberg in his admirable study of 1936 recognises that Brentano had abandoned the use of 'mental inexistence', but notes that he had also dropped the adjective 'intentional' and retained only 'psychic relatedness to the object'. It does not seem however to be correct to say, as Spiegelberg does, that Brentano almost gives up the use of 'intentional' "um nur die mental Immanenz der psychischen Phänomene noch stärker hervortretenzulassen".[14] Brentano in fact did most explicitly reject the conception of 'mental inexistence', and if he also ceased to make use of 'intentional' it must be due to the apprehension that this word was – or had been through his own writings – associated with the idea of mental inexistence. It may also be that when Brentano did speak of 'mental inexistence' he did not mean a sort of literal 'being contained in'.

Just as we are after the integral Husserl, so it is only fair that we should look out for the integral Brentano: in both cases the complete philosophical works of the two authors are just being made available, making possible a reassessment of our usual understanding of the two philosophers.

It has already been emphasised that intentionality cannot any longer be regarded as a distinguishing mark of the mental. For though all that is mental may well be intentional, yet not all that is non-mental is non-intentional. Just as the thesis of intentionality is neutral as against realism and idealism – in fact leads beyond both –, so also it is neutral as against body-mind dualism and behaviourism. What it surely is incompatible with, is a mechanistic interpretation of human behaviour, but not all forms of behaviourism. A philosophy of mind which reduces mental states to bodily behaviour but at the same time replaces an intentionalistic framework for study of behaviour for the Watsonian naturalistic framework is indeed compatible with the intentionality thesis. This point is worth bearing in mind, for it is, or at least used to be, rather rashly supposed that the phenomenologist's interest lay in consciousness and not in bodily behaviour. That this is not so hardly needs to be said today, and this is not true even of Husserl as we shall see in this study. The phenomenologist need not even totally reject the S-R formula, just as Husserl did not totally reject Hume's laws of association. Just as Husserl showed that the concept of association needs to be taken up into the phenomenological concepts of motivation and intentional implication, so we may show that the very concept of stimulus is a covertly intentional concept. Thus suppose there is a mental verb *m* with an intentional object *o*, and suppose its behaviouristic translation into the language of response is *r(o)*. Let us grant also that the translation is adequate, in other words that the mental word *m* is dispensable. But *r(o)* is no less intentional. The truth of the statement 'S makes the response *r(o)*' logically entails neither the existence nor the non-existence of *o*. The response *r* is directed towards *o*, and the sense of this directedness is closer to the mental directedness, and is radically different from the relation of 'being caused by'. To bring in the concept of 'being stimulated by' is of no help, for a stimulus *qua* stimulus is not a cause. Being a stimulus of a response reposes basically upon intentionality.[15]

II

Husserl's concept of intentionality may be said to have developed in several stages though a strict line of demarcation is indeed difficult to draw amongst these. We shall here divide our exposition of Husserl's concept under three heads: the concept of intentionality in the *Logische Untersuchungen*, in the *Ideas*, I, and in the works after the *Ideas*, I.

A. *The Concept of Intentionality in the Logische Untersuchungen*

In the *Logische Untersuchungen*, II.1., we find what has been called by Diemer[16] a static analysis of intentionality, whereas in the next part of the second volume of the same book we are given what may be called a dynamic analysis. This distinction is useful no doubt, but we have to remember that the two do not exclude but rather supplement each other.

The static analysis lays bare the structure of what is called an intentional act whereby the word 'act' has to be taken not in its ordinary usage as meaning an activity or a process, but simply as standing for all intentional mental states. Intentionality is not a kind of activity, but the ordinary concept of action is itself an intentional concept. The distinction that dominates the static analysis and is carried over into the *Ideas*, I but first explicitly abandoned in the lectures on inner time consciousness is that between the '*reell*' component (Husserl does not say 'real' as he is afraid the latter word may suggest the notion of thing-like entities) and the intentional correlate of an intentional experience. By the former he understands all those *Erlebnisse* which themselves are parts of the act under consideration taken as a totality, as an object like any other, but not particularly regarded as an intentional experience. However, the other i.e. the intentional correlate concerns an intentional experience *qua* intentional. But under the intentional correlate, again, Husserl proceeds to distinguish between the intentional object, the intentional matter and the intentional essence of an act. The two judgments '2+2=4' and 'New Delhi is the capital of India' are said to have the same act-quality in so far as both are judgments, but they differ in their intentional matter. The judgment 'There are living beings on Mars' and the question 'Are there living beings on Mars?' differ in their qualities but are said to agree in their intentional matter. Any two acts having the same quality and the same matter may be treated as being identical even if they, regarded as real events in the mental lives of two different persons, may be numerically different. While the quality of an act indicates the peculiar way of its being concerned with the intentional object (its '*Weise der gegenständlichen Beziehung*'), the matter determines its objective direction (its '*bestimmte Richtung auf ein Gegenständliches*'). Thus though the two concepts 'equiangular triangle' and 'equilateral triangle' have the same quality and refer to the same object, yet, since they are not identical representations, we must suppose that the intentional matter in the two cases is different. The intentional essence of an act is next defined as the unity of its quality and matter: this gives us a concept of identity to be applied to the domain of acts for the purpose of phenomenological description.

The sixth Investigation introduces a new concept namely that of the epistemic essence (*erkenntnismäßiges Wesen*) of an act, but this applies only to those acts which are here called 'objectifying acts', which are in fact the theoretical and cognitive acts as distinguished from the affective and the volitional ones. We need not consider here Husserl's theory that only the objectifying acts (which are either nominal or propositional, and in either case either positing or non-positing) are the primary bearers of meaning.[17] For the present, we may mention that while all acts are intentional and possess intentional essence, only objectifying acts possess an epistemic essence. If we remember that the objectifying acts are the cognitive ones, then the above contention that they possess an epistemic essence would appear to be trivial. However, Husserl's thesis here is important: he is by implication saying something more than this apparently trivial thing. He holds that every objectifying act necessarily has an intuitive basis. The nature and the rôle of this intuitive basis surely differs in the case of the different kinds of objectifying acts. It may, when the act is symbolic, serve as a mere sign of the intentional matter; in the case of knowledge in the strict sense it may serve to fulfil the meaning intention. While the intentional essence of an act was defined as the unity of the act-quality and act-matter, the epistemic essence is now defined as the unity of quality, matter and the intuitive content. Just as the concept of intentional essence provides us with a concept of identity of acts in general, so the concept of epistemic essence enables us to speak of *the same* knowledge, howsoever this knowledge may otherwise differ.

This static analysis of intentional act is supplemented by a dynamic analysis in the sixth Investigation. This new point of view brings to the forefront *the dialectics of intention and fulfilment.* This dialectics shows that the intentional act involves an aiming at, a project, an active thrust demanding actualisation, fulfilment or satisfaction.[18] In it we begin to gain two fresh insights, both definite advances: for one thing, the minimum sense, or rather the central core of temporality involved in intentionality, in fact the essential relationship between the two concepts, begins to reveal itself before our eyes; for another, the early phenomenological notion of absolute givenness, of total intuitive grasp, makes room for the ever receding goal of absolute fulfilment or complete coincidence of the originally empty intention with intuitive fulfilment. Indeed though the spectre of a total coincidence persists in the *Logische Untersuchungen* where 'knowledge' is still defined in terms of it, yet Husserl soon abandons it as a description of what is ever happening when we are said to know, but continues to retain it as an ideal towards which our intentional life asymptotically aims. We learn both that the *presence* has to be measured by the sense i.e. the intention it fulfils, and also that the sense, the intention, has to be appraised by evidence, every claim to be measured by 'seeing'.[19] What is of still greater contemporary relevance is that Husserl goes beyond the positivistic restriction of all evidence to the sensuous, and leaves room open for different kinds of fulfilment and for different strata of verification.

B. *The Concept of Intentionality in the Ideas I*

The structure of the intentional act as revealed by the static analysis is basically retained in the *Ideas* I. The *reelle* components of the act are now grouped under two heads: the *hyle* and the *noesis.* As contrasted with them, the intentional correlate is now called the *noema.* Just as with the concept of the *noema* we get into the real heart of the concept of intentionality, so with the concept of the *hyle* we face one of the chief bugbears of the later day phenomenologists. Many have found in the latter a trace of uncritical sensualistic empiricism, others have found in it a survival of the Aristotelian-Kantian hylo-morphic scheme. Whatever may be the value and the historical affiliation of this concept of *hyle*, the fact remains that it is there and also that Husserl himself came to suspect it later on, and it is doubtful if he was ever able to do away with it entirely.*

While the *noema* as the sense (as distinguished from the intentional object which is the reference: Husserl uses 'meaning' for both) is both subjective and objective, "*ein Zwitter subjektiver Immanenz und transzendenter Objektivität*",[20] the *noesis* is both psychological (being a real component of an act) and transcendental (being meaning-conferring) at the same time. Their correlation constitutes the wonder of wonders, one of Husserl's most significant discoveries. But even this correlation (succinctly summed up in Husserl's sentence: "No noetic phase without a noematic phase that belongs specifically to it"[21]) has various aspects which may be distinguished as follows:

(a) Each act has its own intended object.
(b) Each act has its own *noema.*
(c) Different acts may have the same *noema.*
(d) Different *noemata* may refer to the same object.
(e) Each act-phase has its noematic-phase.

In any case, at the noematic level there is achieved a unification which raises the *noema* above the temporality and particularity of the noetic acts, but which also brings to light the fact that the noetic act is not the merely psychological, temporal event that it at first appears to be but is also psychological-transcendental in one.

While the *irreal noema* stands over against a multiplicity of real acts, a multiplicity of *noemata* refers, through passive synthesis of identification, to an identical object, real or ideal. Thus the *irreal noema* mediates between the real intentional act and the real (or ideal) intended object. In relation to the acts the *noema* is an identity, in relation to the noematic multiplicity the intentional object is an identity. To put it in the language of the later phenomenology of constitution, the constitution of the intended object by its correlative noetic acts is mediated by the constitution of the *noema.*

It would be an error to suppose, as Gurwitsch does, that the noetic-noematic correlation constitutes the central core of Husserl's notion of intentionality.[22] The relation of *noesis* to *noema* is not the same as the relation of consciousness to its object.[23] The reference of consciousness to its object is *made possible* through (a) the correlation and (b) the noematic 'nucleus' by virtue of which the noema also refers to the object, and (c) the consequent noematic intentionality. It should also be borne in mind that the correlation characterises only the act-intentionalities, and cannot be meaningfully extended to intentionalities other than acts. Thus for an adequate understanding of Husserl's concept of intentionality we need to go beyond the correlation thesis.

C. *The Concept of Intentionality in the works published after the Ideas I*

There is no doubt that the two primitive notions of the *Ideas* I, the noetic act and the *hyle*, lose their privileged status in the lectures on inner time consciousness: they also are now shown to have been constituted in the inner time consciousness. The act-matter scheme, the hylo-morphic framework dating back to the *Logische Untersuchungen* now collapses. Acts and data are also immanent objects, they are possible objects of inner reflection. They are temporal unities. As such they both are to be analysed into primal impressions (*Urimpressionen*), the *nows* with their retentional and protentional horizon. Husserl, in these lectures, seems to be oscillating between a real and an intentional analysis. When, for example, he decomposes a total act into partial acts or act-phases – the tendency is there in the *Ideas* I also – the analysis seems more *reell* than intentional; when the total act-intention is decomposed into partial intentions which remain anonymous behind it, the analysis is intentional. The former decomposition may be misleading in so far as an act may be regarded as having no temporal stretch at all, it may be concentrated in one moment. The second decomposition would still hold good of it: the partial intentions would not break up this momentary character of the total intention. The partial intentions are not phases of the act, but its media, conditions of its possibility. As the act is to be traced back into partial intentions that are not acts, so the supposed datum, the unmeaning stuff, the sensory manifold – as a temporally enduring object – is resolved into momentary phases each with its retentional-protentional horizon. However it should be borne in mind that when the *datum* is said to have an intentional structure, that is so in a sense radically different from that in which acts are intentional. Its intentionality is simply its temporality, it is not intentional in the sense of being *of* anything.[24]

It has been pointed out[25] that this formal analysis fails to account for the material i.e. the contentual aspect of the hyletic datum. It would then seem that even if we discard an absolute dualism between form and matter, we cannot do away, at any given stage, with a dualism relative to that stage. The phenomenologist has to retain this relative dualism without succumbing to an easy sensualistic empiricism with regard to the *hyle*, and without identifiying all intentions with acts. This indeed is a problem for both Husserl and the later day phenomenologists.

Intentionality as Constitution

We are thus led from the conception of intentionality as a self-transcending reference to the conception of it as noetic-noematic correlation, and then from the latter to the concept of it as constitution. The last conception, anticipated in the *Ideas* I and also in the lectures on time consciousness, receives much extension in the later works. In fact we may distinguish between two senses of 'constitution': constitution as *Sinngebung* and constitution as transcendental production. Let us call these 'constitution$_1$' and 'constitution$_2$'. The former is a consequence of the doctrine of correlation, it presupposes the hylo-morphic scheme, and it applies only to act-intentionalities and so valid only within a limited perspective.[26] Fink holds constitution$_1$ to be indeterminate (*unbestimmt*), neither fully receptive nor fully creative, and constitution$_2$ to be productive and creative. However while it is true that Husserl came both to suspect and give up his earlier hylo-morphic scheme, we should also remember that he never overcame his deep distrust of the notion of *creativity* of consciousness.[27] One cannot but agree with Berger that the notions of 'production' and 'activity' are mundane conceptions and do not apply as such to the transcendentally reduced consciousness. One has also to remember Ricoeur's insightful remark:

> What interests Husserl in consciousness is not its genius, its power to invent in every sense of the term, but the stable, unified significations into which it moves and becomes established ... Phenomenology is a philosophy of "sense" more than a philosophy of freedom.[28]

We shall point out, in a later context, the rather strange fact that with the development of the doctrine of constitution Husserl also came to abandon the idea of total bodily givenness of objects and that of complete transparence of consciousness to itself: this also corroborates our thesis that even constitution$_2$ does not amount to creation or production.

But the transition to constitution$_2$ is made possible through several lines of thought: first, the discovery of the intentional structure of the hyletic stratum; then, the exhibition of the constitution of the acts; the resulting abandonment of the hylo-morphic scheme; the eventual coming into prominence of the intentionalities other than acts, and emergence of a host of new concepts like those of 'genetic constitution', 'passive synthesis', 'pre-predicative intentionalities', 'horizon intentionalities', 'unconscious intentionalities', 'anonymous intentionality' and 'operative intentionality'. These concepts not only lead to the concept of constitution$_2$ but also make possible the very different philosophies of the later day phenomenologists like Merleau-Ponty. They deserve therefore our careful attention.

The idea of 'passive' synthesis is already to be met with in the *Ideas* I, but in the form of that (*passive*) synthesis of identification by which various *noemata* coalesce so as to make reference to one identical object possible. The *Ideas* I also contains the idea of horizon as what is in an improper sense given with all things (*uneigentliche Mitgegebenheit*) but with more or less indeterminateness to be determined in accordance with a pre-given style (*Bestimmbarkeit eines fest vorgeschriebenen Stils*)[29]; also the idea of the threefold temporal horizon of every experience[30], worked out in greater detail in the lectures on time consciousness. We also meet there with the distinction between theoretical (doxic) and not-theoretical acts[31], the former being held to be the basis of the latter; but this distinction is not quite the same as that between act-intentionalities and intentionalities other than acts. The primacy of the theoretical acts persists in the *Ideas* I.[32]

The *Ideas* II proceeds to found theoretical acts on "*vorgebende*" intentionalities which are not themselves theoretical.[33] But the specific sense of an operative intentionality is still lacking.

The overall importance of 'passive synthesis' comes to light only later on, first in the lectures of 1918–26[34], then in the *Formale und Transzendentale Logik* and in the *Cartesian Meditations*. The *Cartesian Meditations* distinguish between active production of abstract entities like collections, numbers or inference-structures and passive constitution of pre-given things like the physical objects which confront us in life as ready-made.[35] A reconstruction of the sense of their reality, of their pre-givenness, of their ready-made character would reveal their constitution: in this constitution, association no doubt plays a major rôle but so does anticipation as well. In fact, whatever else Husserl may be doing in his analysis of the passive constitution of pre-given things, his ground had been covered partly by Kant's notion of the transcendental synthesis of imagination and partly by the more modern notion of logical construction. The *Phänomenologische Psychologie* regards the passive synthesis as belonging to experience itself and as forming "*die Unterlage für das Ins-Spiel-setzen der Aktivitäten des beziehenden, logisch verallgemeinernde Allgemeinbegriffe und -sätze konstituierenden Tuns*".[36]

'Passive synthesis' is also called in the *Cartesian Meditations* 'passive genesis'. The typically new element in the notion of genetic constitution is the element of historic achievement. Thus in the *Formale und Transzendentale Logik*, referring to judgments and their constitution, Husserl writes:

> Es ist eben die Wesenseigenheit solcher Produkte, dass sie Sinne sind, die als Sinnesimplikat ihrer Genesis eine Art Historizität in sich tragen ... dass man also jedes Sinngebilde nach seiner ihm *wesensmäßigen Sinnesgeschichte* befragen kann.[37]

To uncover the genetic constitution is to lay bare these sedimentations of meaning, "to reactivate the original encounter".[38] The idea of hidden achievements which have to be uncovered by phenomenological reflection is not far away.

The idea of genetic constitution applies to the pre-given real things and to the ego.[39] It may even be said that the cultural objects also have a genetic, but not surely passive constitution; physical objects are passively and genetically constituted but not historical achievements; the ego is constituted by *habitus*, through a sort of auto-genesis, though it never quite becomes an object but at best somewhat like an object. From this we may conclude that not all genetic constitution is passive, though all passive constitution is genetic. Further it would be fair to add that not all genetic constitution is historical achievement.[40]

The specific sense of horizon intentionality[41] also comes gradually to the forefront in the late twenties. The *Cartesian Meditations* refers to the horizon as the objective sense, implicitly meant and foreshadowed in the actual cogito; the predelineation is never perfect, but in spite of its indeterminateness it does have a determinate structure.[42] The object as a *pole of identity* is meant "expectantly" through the "horizon intentionalities".

The ideas of unconscious and anonymous intentionalities emerge in various contexts. The intentionalities which passively constitute the pregiven world are "anonymously" there; the experiencing subject knows nothing of them.[43] Phenomenology has to rescue them from this anonymity. It penetrates the anonymous intentional life and uncovers its synthetic processes.[44]

It would be an error to equate the anonymous intentionalities with either the unconscious ones or the intentionalities other than acts. Thus, e.g. the horizon intentionalities are not acts but are neither anonymous nor unconscious. Acts may have contributed to the sedimented sense of a pre-given object, and in such a case it is these acts which have become anonymous. Unconscious intentionalities, on their part, may be either acts or not. There are in fact unconscious desires, love and hatred which are acts. The notion of unconscious intentionality makes its appearance in course of Husserl's attempt to take account of the findings of modern depth psychology.[45]

The concept of operative intentionality (*fungierende Intentionalität*) found only in the form of meagre hints in Husserl's later writings has been made much of by later phenomenologists. In fact, one may say that all the above concepts of intentionality – the concepts of passive synthesis, genetic constitution, horizon intentionality, anonymous intentionality and unconscious intentionality – are brought together in the last papers under the title 'operative intentionality'. Already in notes to the manuscript of the *Erste Philosophie*, Part II[46], we are told of the implicit "*Verflochtenheit aller Intentionalität in einem Leben*". A note of 1925 to the *Phänomenologische Psychologie*[47] speaks of the intentional life as *Weltleben*. The adjective '*fungierende*' is used, along with '*lebendig*', perhaps for the first time in the *Formale und Transzendentale Logik:*

> Die lebendige Intentionalität trägt mich, zeichnet vor, bestimmt mich praktisch in meinem ganzen Verhalten, auch in meinem natürlich denkenden, ob Sein oder Schein ergebenden, mag sie auch als lebendig fungierende unthematisch, unenthüllt und somit meinem Wissen entzogen sein.[48]

The operative intentionality is here said to be that mode of the intentional life which remains unthematic, undisclosed and thus removed beyond my knowledge although it carries me forward in my theoretical as well as practical relationship. To say that operative intentionality "*ist* nicht schlechthin, sondern sie fungiert"[49] is hardly enough, for the same may also be said of all, even act intentionalities. The operations of the so-called operative intentionalities are anonymous, they are not made thematic except to phenomenological reflection. They are not acts and also not cognitive. Yet they confer sense on the pre-given world, and thus provide the basis for further creation of significance. According to Fink, it is "die lebendig sinnbildende, sinnleistende, sinnverwaltende Funktion des Bewußtseins, welche zu den einfachen seelischen Einheiten der Akte sich selbst verdeckend zusammenschliesst".[50] According to Merleau-Ponty, it is "that which produces the natural and ante-predicative unity of the world and of our life".[51]

It appears from the above that the notion of operative intentionality is a logical consequence of the entire trend of thought we have been sketching in the foregoing pages. It is the last ground and source of all meaning and significance. How precisely this notion is related to two other notions appearing and playing a major rôle in Husserl's later thought: namely, the notions of *Lebenswelt* and *Welterfahrendesleben*, is indeed a most difficult problem for an exegesis of the later Husserl.[52]

III

The existential phenomenologists like Heidegger, Sartre and Merleau-Ponty have made much use of the concept of intentionality, but they have subjected Husserl's various theses to searching criticisms and have very often radically altered them. Their main criticisms of Husserl concern several different issues. Some like Sartre have been concerned with the question of the transcendental ego as the *source* of intentional acts; others like Heidegger have sought to rehabilitate intentionality as an ontic or even as an ontic-ontological relation, and consequently do away with the concept of *noema*. Most of them have found fault with the Husserlian concept of the *hyle* and the consequent hylo-morphic scheme, though all these critics surely differ in their attitudes towards the doctrine of constitution in any of its two forms. Another central issue has been the question about the *locus* of intentionality, and finally the problem regarding the relationship between the concepts of transparence and intentionality have been made to bear on all these discussions. Let us take a close look at each of these points of controversy.

1. Sartre, as is well known, rejected Husserl's conception of the transcendental ego and defended the non-egological concept of intentionality of the first edition of the *Logische Untersuchungen*.[53] First, the ego is not needed to account for the unity of consciousness, intentionality is adequate to unify consciousness as Husserl himself has shown in his lectures on time consciousness. Again, locating the ego *within* consciousness would be to destroy the translucence of consciousness: the 'I' would be an opaque centre which consciousness by its very nature would reject. Furthermore, the unreflective consciousness does not have the form 'I think', though it is surely transparent to itself. The 'I' is an *object of* consciousness, it is *constituted* by reflection, and is not the source of the constituting function.

2. The idea of noetic-noematic correlation is in general abandoned[54], and intentionality is sought to be understood as an ontic relation. Phenomenological reduction which, as we saw, is a necessary methodological preliminary for the Husserlian notion is rejected, for existentialism is to be a phenomenological *ontology*. Intentionality becomes, as with Heidegger, an openness to Being, but also being-in-the-world of the *Dasein*. The world, Sartre tells us, is not a *noema* but the in-itself.[55] Intentionality is characterised as transcendence[56], which Husserl's intentionality – according to Sartre – is not.[57] It is transcendence in a two-fold sense: it refers to a being outside of itself, and it transcends the ontic being towards the ontological i.e. the meaning of being. For Merleau-Ponty also the primary intentionality is the body's motility and so is an ontic relation directed towards the world, and not towards an *irreal noema.*

3. The concept of *hyle*, as is to be expected, comes in for sharp criticism. Sartre has argued that though Husserl came to the recognition of the hyletic component of consciousness in order to account for the element of passivity in knowledge, yet the doctrine itself creates more difficulties than it solves. "The *hyle* in fact could not be consciousness, for it would disappear in translucency and could not offer that resisting basis of impressions which must be surpassed towards the object."[58] This objection is genuine, provided only that we accept Sartre's basic thesis that consciousness is always and wholly transparent. Again, even if we do admit the *hyle*, it does not facilitate the self-transcending function of consciousness, we still fail to understand this subjective layer towards objectivity. Following Heidegger, Landgrebe[59] has attempted to get rid of the atomistic and sensationalistic conception of sensation common to both Kant and Husserl by construing 'sensing' as "a structure of the Being-in-the-world". The many facts to which Landgrebe appeals (e.g. the fact that sensation is closely connected with movement) are undeniable, but it is still very doubtful if the general notion of Being-in-the-world is adequate to account for the particularity and variety of sensuous data. It has also been pointed out that sensations have properties, like the spatial properties, which make them unfit for being components of consciousness; they, in fact, are intimately connected with body.[60] We seem to be back with the central issue of the status of the body in Husserl's thought. So far as Merleau-Ponty is concerned, it is only to be expected that he will also join with these critics of the doctrine of the *hyle*. We have to remember his original criticisms of the very concept of sensation in the opening chapters of the *Phenomenology of Perception*. But what is more, all sense experience is for him already intentional and charged with significance. Thus he writes:

> Every sensation is already pregnant with a meaning inserted into a configuration, and there is no sense-datum which remains unchanged from illusion to truth.[61]

4. Husserl's doctrine of constitution has aroused quite different responses amongst the existentialist phenomenologists. In Heidegger, with the increasing emphasis on the concept of *openness* of *Dasein* to Being, the notion of constitution$_2$ almost drops out; though the notion of constitution$_1$ dimly survives in so far as the project of *Dasein*, its *Verstehen* understood as the prospective dimension of being-in-the-world – which is Heidegger's reconstruction of the Husserlian notion of the noetic act[62] – confers meaning and significance on the other, and also in so far as *Befindlichkeit* – again Heidegger's reconstruction of the Husserlian *hyle* – reveals the pre-given accomplished facticity of *Dasein*, the original situation of having been 'thrown' into the world. Sartre more explicitly rejects the formulation of intentionality as a constituting function,[63] and gives two reasons why this should not be so. In the first place, for consciousness to be of something is to be confronted with a concrete and full presence which is not consciousness. But also, the supposed constitution of the object would involve an infinity which could not possibly be given, whereas there could be nothing in consciousness that is not immediately transparent to it.[64] It is in Merleau-Ponty that the doctrine of constitution is rehabilitated but fully freed from the idealistic overtones of Husserl's thought. As we have already said, he rejects the notion of the *hyle*, but the notion of constitution$_1$ nevertheless survives. For, now the *Sinngebung* is assigned to the body which performs the function of Husserl's noetic act. The body confers meaning on the world, but this is not arbitrary but reveals, as it were, a body-world pact[65] so that the body's worldfinding activity grounds the general structure of the world. The Husserlian conception of an irreal noema is not there, but the idea of correlation of the reciprocity of the elements entering into the structure of intentionality is given a new interpretation. The doctrine of constitution has to take care of the fact that "the unity of the world, before being posited by knowledge in a specific act of identification, is 'lived' as ready-made or already there"[66]; as well as of the fact that if the world were constituted in an absolute self-consciousness then the world would have been totally transparent to consciousness, there would have been no error, no illusion, no opacity in the world as well as in consciousness. Consciousness therefore does not embrace or possess the world, but itself a project of the world, is yet perpetually directed towards the world, giving it sense and significance. The latter Husserlian notion of operative intentionality is used to care of the first of the above-mentioned two facts; and a doctrine of degrees of transparence[67] as well as of degrees of intentionality, combined with the thesis that it is the body which exercises the basic constituting function, which enables Merleau-Ponty to take care of the latter fact. But this giving of sense is a continuous process. We give sense, not a ready-made sense from a transcendental pedestal (as with Kant or Husserl) but always in degrees through our continuous explorations. The fully determinate object with completed sense and the absolutely transparent constituting consciousness are both rejected: a consequence of taking time and history seriously! The motion of operative intentionality with its anonymous mode of working leads to the notion of forgetfulness: the task of phenomenology is to make explicit this 'cryptomechanism", to catch it in the very process of conferring sense and so "to foil its trick"[68] by capturing the process of emergence of definiteness and meaningfulness. All this hinges, as should be obvious, on three basic propositions:

(a) Intentionality is always a matter of degree.
(b) Transparence is always a matter of degree.
(c) The basic intentionality is that of the body,[69] which gives the

> lie to the absolute opposition set up by Sartre between the for-itself and the in-itself: the body is intentional, and is not yet fully transparent.

5. The existential phenomenologists also differ from Husserl with regard to the *locus* of intentionality, i.e. with regard to the question, 'What sorts of things are intentional?'. As we have noted before, the Brentano thesis about the alleged criterion of the mental has been left far behind. For Husserl, nevertheless, consciousness is a locus of intentionality; if it is the only thing which possesses intentionality remains yet to be seen. Sartre in this respect is an orthodox Husserlian, for with him too consciousness alone is intentional. The great dualism between the for-itself with its complete self-transcendence and total transparence – both entailing its nothingness – and the in-itself dominates his ontology, and the status of the body remains a problem. Heidegger explicitly rejects the primacy of the concept of consciousness; the dualism of consciousness and Being, a Cartesian ghost haunting the *Ideas* I, is sought to be exorcised. Intentionality, whether in the sense of openness to Being or in the sense of being-in-the-world characterises *Dasein*, not consciousness. While this change persists in the later works, the primacy accorded to intentionality in the *Sein und Zeit* is soon recanted: it is sought to be derived from the rather metaphysical notion of historicity and destiny of Being and its world-project. Intentionality is *made possible*, we are now told[70], by transcendence and not *vice versa*. It is understandable that Sartre should have found Heidegger's concept of *Dasein* unsatisfactory in so far as it excludes consciousness[71]: Sartre's for-itself is consciousness and not *Dasein*. While Sartre goes back to the early Husserlian thesis that intentionality belongs to consciousness and consciousness alone, Merleau-Ponty's entire philosophy marks a return from the pure for-itself to concrete human existence; but unlike Heidegger, intentionality is located in the body, in that ambiguous region which refuses to be subsumed under any member of the Sartrean pair.

In all these philosophers there is a tendency towards what may be called a monistic conception of intentionality, according to which intentionality *originally* belongs to one and only one sort of things, and if there are other sorts of things which also exhibit it their intentionality must be derivative from that one original intentionality. In other words, there is only *one original* ascription of intentionality (whether to consciousness, or for-itself, or *Dasein*, or body); and of course all other ascriptions are derivative from, and presuppose, this one. This monistic tendency is relieved by Merleau-Ponty's insistence on the degrees of intentional function, and it is well worth consideration if this clue may be utilised to overcome the monistic tendency and to accord recognition to *radically different* types of intentionality.

6. An issue which emerges out of the foregoing discussion may be formulated as follows: how are the two features of consciousness, namely its transparence and its intentionality, related? Is there a sort of opposition between them, so that in fact consciousness can only possess one of these two and not both? Or, in case there is no such opposition, should we say that any one of the two is primary and the other derivative?

Now, that the transparence of consciousness – or, what the Indian philosophers called its '*svayaṃprakāśatva*' – is incompatible with its supposed intentionality is most forcefully defended by the great Indian idealist Sankara: for him, the self-illuminating consciousness cannot be *really* of an object, so that the intentional reference is only apparent – the product of a metaphysical ignorance! Now this is not the place to enter into a detailed examination of this view. But Indian thought also contains an exactly opposed answer: namely that since consciousness is always intentional i.e. is of an object, it cannot at the same time be of itself. It would then follow – as was in fact held by the school of the Nayiyaikas – that the intentional consciousness is never transparent, it becomes aware of itself only through a subsequent act of introspection. At the same time, there is a third trend of thought – upheld by Ramanuja – according to which consciousness is always both: it is immediately aware of itself at the same time as it is also of an object. Now this precisely is what would seem to be the view of Sartre. But ascribing to consciousness total transparence, total coincidence with itself would rule out the possibility of all error, illusion and self-deception. One may even suspect that even for Merleau-Ponty there is an opposition between total transparance and total intentional absorption. Consequently he is led to deny both: neither the *cogito* nor the *cogitatum* is grasped fully. There are in fact, as we have already emphasised, only degrees of both.

While Merleau-Ponty's concept of degrees of intentionality and transparence is highly valuable and almost indispensable for any phenomenology of consciousness, we must nevertheless say that the supposed incompatibility between the two functions does not hold good. The two do not vary in inverse proportion. On the contrary, as Paul Ricoeur has so aptly shown, sometimes the reverse may well be the case: there are acts in which greater self-awareness is attained with increasing absorption in the intentional object, as in the case of acts of decision and resolution.[72]

If there is no opposition between the two, one may ask, as we have above, is any one of the two more primitive, is any one dependent on the other? Now, it would seem that if we have to make room for the supposed unconscious intentionalities we must have to recognise that intentionality is independent of transparence. But then it would seem that the transparence of consciousness depends on its intentionality. This may seem to have been the view of Sartre: consciousness, on this view, is translucent only in so far as its total intentionality leaves nothing within it, empties it of all contents, and reduces it to nothing as it were. It is this emptied consciousness which is wholly transparent. True, as the Nayiyaikas contend, if consciousness be wholly directed towards an other it cannot at the same time be directed towards itself. But this self-givenness without being directed towards itself, this non-positing, pre-reflective self-givenness of consciousness is exactly what we have called its transparence; and it could be said that intentionality constitutes a necessary, if not a sufficient condition, of its possibility.

IV

In the preceding section we have briefly touched upon the main issues emerging out of the criticisms of Husserl's concept of intentionality by the existential phenomenologists. It remains to be seen to what extent these points may be taken care of from the point of view of Husserlian phenomenology, especially in view of Husserl's later writings.

1. Sartre's criticism of Husserl's concept of the transcendental ego as the source of intentionality may be met in either of two ways. One may argue[73] that Sartre is right in insisting that the ego is not empirically given and that the *mine* is capable of being accounted for within the framework of a non-egological theory of consciousness, but at the same time one may contend that this empirically given intentional structure *presupposes* the transcendental ego. Now this solution, it seems to me, is not Husserlian. The ego that is merely the presupposition is 'transcendental' in the Kantian sense but not in Husserl's sense of that word. Husserl however did get beyond the particular version of the doctrine of the ego which Sartre criticised. For, he came to recognise that even the transcendental ego is constituted though in a manner radically different from the way objects are constituted. In the *Ideen*, II, Husserl says that the pure ego is not in need of constitution[74], but in a *Beilage* to it[75] he adds that "schon vorher muß die Lehre vom reinen Ich – zunächst als Pol – revidiert(?) werden"; and expresses the view that though the ego is a pole of identity, it is an "unselbständiges Zentrum für Affektionen und Aktionen". Of course, the ego is not to be found in the stream of experiences – and so far Sartre's point is not disputed.[76] We are even told immediately after that "Die absolute Identität erkenne ich in der Reflexion". However it is in the *Cartesian Meditations* that we come to the explicit formulation of the doctrine of the self-constitution of the pure ego through habitualities.[77] In fact, it is important to remember the distinction between transcendental ego and transcendental subjectivity.[78] The latter alone is premundane, pre-objective, meaning-giving consciousness; the former is a constituted unity, though its mode of constitution is radically different from the mode of constitution of objects. However, this development of Husserl's thought renders the whole thrust of Sartre's criticism pointless.

2. The central core of the second set of objections also is based on a misunderstanding of Husserl. By positing the irreal *noema*, Husserl did not want to deny that intentionality relates to beings, or even to Being (whatever that may mean). As we have emphasised in course of our exposition of Husserl's concept of intentionality, he was clearly aware of the fact that the noetic-noematic correlation was not the same as – nor could it be construed as a phenomenological reconstruction of – the intentional reference to object. The *noema* is not the intended object. But such is human consciousness, one could say, that its intention posits the *noema* and through it as it were reaches out to the object beyond. But even this way of putting the matter is misleading. For, the *noema* and the object belong to two radically different dimensions – linguistically to two totally different universes of discourse, phenomenologically to two radically different modes of givenness. The possibility of turning away from the object-oriented attitude to the subsequent noema-oriented attitude is an inherent possibility of consciousness. But the introduction of the *noema* should not be regarded as shutting the door for ontological encounter. It may be supposed that Husserl is introducing a form of the content theory. Possibly he is doing that. But a content theory should take into account two facts. For one, the content is never given together with the object: the two are given in two radically different attitudes. This radical difference in their modes of givenness bears testimony to the radical difference in their modes of being. For another, the content is not another object standing in between the mind and the object, obstructing the reference by its opacity. It is rather a transparent medium which makes intentional reference possible. In defending the doctrine of the *noema* we have said that perhaps human consciousness is such that its intentionality posits such an irreal medium of reference. Hasn't Merleau-Ponty said that we are condemned to meanings? And yet we are in rapport with the world! It is true that the intentionality thesis requires us to reject the conception of a representative consciousness. But the *noema* is not a representative, not a Lockean Idea, not a copy of the transcendent object. Sartre's for-itself, the totally empty consciousness, fully open to Being but emptied of all contents, the consciousness that is out and out nothing – this precisely is not the human consciousness. A theory of consciousness that would be adequate and fair to human consciousness should of course (i) take into account its intentionality and therefore the fact that it is not a representative consciousness; but (ii) it should also recognise that man is nevertheless condemned to meaning, that it is not passive openness but also active *Sinngebung*. The concept of *noema* is an attempt to take both these facts into consideration.

3. The much belaboured doctrine of the *hyle* had had very few defenders amongst Husserl's followers. It is also true that Husserl himself has disowned the version of the *Ideas* I as having been due to the influence of a sensualistic empiricism. But it cannot also be straightaway denied that some form of the doctrine has to be recognised by any phenomenology of consciousness. Sartre's fear that to locate the *hyle* within consciousness would amount to inserting into the all-transparent consciousness a "centre of opacity" would be justified only if we assume that consciousness is in fact all-transparent. But this is an assumption which one may very well question. The points raised by Merleau-Ponty and Gurwitsch are on surer footing: but they are directed against an atomistic and sensualistic conception of *hyle*. The supposed discrete sensation is quite possibly a myth – a product of theory and not a datum of consciousness. The most elementary data may exhibit intentional structures of various sorts. Nevertheless the basal thesis that there is some kind of impressional matter does not stand or fall with the conception of sensation; and it is in the former form that the doctrine of *hyle* may still survive.[79] Furthermore, as Ricoeur has shown[80], this doctrine may be indispensable for those phenomenologists who would like to take the psycho-analytical concept of the unconscious more seriously.

Husserl however came to emphasise the connection between the hyletic stratum of our experience with the body. In the *Ideas* I, where the *hyle* is regarded as a real component of consciousness, the body is regarded merely as the connecting link between pure consciousness – which is by itself an absolute – and the real world.[81] He does recognise there, however, that states like a state of pleasurable feeling have an intentional reference to the 'Menschen-Ich' and the human body, but adds that when abstraction is made from this reference the state concerned undergoes an alteration such that it finds place in the pure consciousness and loses all natural significance.[82] It is this abstraction from what would otherwise seem to be an essential bodily reference which leads him to locate all *hyle* directly within pure consciousness as its real component. The *Ideas* II contains Husserl's first systematic phenomenology of body.[83] We need not consider here all the different levels of understanding body which Husserl elaborates: body as the 'zero-point of orientation"[84], body as the field of localisation[85], body as the organ of will[86] etc.... What interests us here is the clear realisation on the part of Husserl that a large part of the hyletic stratum (e.g. touch) is directly founded in the body and that, through it, the intentional component (e.g. the touching) also receives an indirect localisation.[87] This localisation entails for Husserl[88] that the body is already more than a material thing and possesses a psychic layer ("es hat schon eine zum 'Seelischen' gehörige Schicht") – so that when we speak of the physical body we are abstracting from the total body. He even comes to call the body "das subjektive Objekt".[89]

We may therefore conclude (i) that the concept of *hyle* is indispensable to account for the concrete fullness of the content of our human conscious life; (ii) that it may and should be freed from the atomistic-sensualistic form which the *Ideas* I gave it; (iii) that the conception is also needed to take into account the psycho-analytic concept of the unconscious within the framework of phenomenology; and (iv) that the concept needs to be founded on an adequate phenomenology of body and its psychic stratum. In all these respects, the later writings of Husserl contain valuable suggestions.

4. Some of the criticisms made against the interpretation of intentionality as a constituting function are not sustainable. That the supposed constitution involves an infinity will indeed be admitted by Husserl[90]: there is always a presumption, an element of hypothesis, an 'I can go on...' involved in it. The constitution of real things is characterised precisely by this presumptive character. Sartre complains that if this infinity were there it would have been transparent. Here again he is assuming that total transparence is an accomplished fact. Were it so, reflection would have been rendered superfluous. It is reflection which discovers this 'I can go on', this infinity involved in the synthesis of an endless multiplicity of perspectives. The infinite series need not be completed. A phenomenology of constitution does not *fully* lay bare the constitution: it on the other hand courts the tragedy of its own failure to complete the infinite series. But this subjective failure has its correlate in the objective essence of the constituted thing as well. Once we understand this, Merleau-Ponty's charge that a world constituted in consciousness would stand fully revealed loses its biting force. The charge cannot be sustained on the basis of Merleau-Ponty's own concept of consciousness as ever possessing only a degree of transparence.

Whereas the *Ideas* I restricts the constituting function to the pure consciousness and its intentionalities, the two other volumes of the *Ideas* mark an advance in this respect. With the growing awareness of the importance of the body and with the growing liberation of his conception of body from the naturalistic framework, Husserl also comes to realise that the constitution of both nature and psyche is essentially bound up with the constitution of the body. The *Phänomenologische Psychologie* carries this a step further ahead: here Husserl, possibly for the first time, speaks of 'bodily intentionality', and says that the intentionalities in which things come to be given cannot be investigated without investigating the intentionalities of one's own body (*"eine verantwortliche wissenschaftliche Aufwicklung der beiderseitig aufeinander bezogenen Intentionalitäten"*).[91]

5. We have just noted that Husserl came to speak of bodily intentionalities. However, we do not notice in him either of the two possible monistic moves which are likely at this stage. He does not say that bodily intentionality is but a pale reflection of the intentionality of consciousness, that the body only appears to be intentional owing to the intentional consciousness which inhabits it. He also does not say that the basic intentionalitity is that of the body, and that the intentionality of consciousness is only an epiphenomenon of it. That he does not take to any of these two likely monistic moves shows that descriptive phenomenology does not yield ground even in his later writings to speculative system building. We are thus led to the recognition of radically different sorts of intentionality: the precise relationship between these is a matter of descriptive investigation rather than of *a priori* metaphysical speculation. In any case the body mind dualism reduces to a dualism of intentional function.

6. As we have emphasised before, the issues we have been discussing lead us to the question of the transparence of consciousness. Husserl never held the view that consciousness is totally transparent. On the other hand, he did believe that reflection can bring to givenness what is there in a conscious act. There is something deceptive about this view, no doubt. It has been held by many philosophers (by the Nayiyāyikas, for example, amongst Indian philosophers) that just as the table in front of me is apprehended by an outer perception, so a conscious state within me is for the first time apprehended by an inner perception. I do not think Husserl ever held such a view which is obviously mistaken in relaying too heavily on the analogy between the givenness of outer objects and the mode of givenness of consciousness. Husserl of course did have the tendency to construe the latter as another higher order intentional act directed towards the unreflective act, but he was also aware that the talk of 'inner perception' in this connection might mislead.[92] All experiences, he says, especially all intentional acts are conscious not only in the sense that they are *of* something but also in the sense that they are in a sense, which does not apply to external objects, *wahrnehmungsbereit*: so that when we do apprehend them in reflection we also apprehend them as always having been there. Now even this concession does not suffice to bring out that peculiar mode of givenness of an *Erlebnis* prior to its being reflected upon which Sartre calls its prereflective transparence. In fact, Husserl never came to distinguish between reflection (which is a higher order intentional act directed towards the act being reflected upon) and reflexivity (which is the peculiar mode in which every conscious state is given to itself even prior to its being reflected upon). Every *Erlebnis*, one may say, is reflexive but not reflected upon. Every intentional act is not an 'I think', true; but every one of these is surely reflexive in the sense of having some degree of self-awareness not amounting to reflection. Thus the reflexivity of consciousness (which the Indian philosophers called its *svayaṃprakāśatva*) cannot be reduced to its intentionality. But one may very well argue, *a la* Sartre, that *only because* consciousness is intentional and *in so far* is free from any opaque content within it, it is transparent.

If this be so, we may try to give some sense to locutions like 'degrees of transparency' or 'degrees of intentionality', and we may also in the same connection give some justification for some form of the doctrine of *hyle*. The argument may be formulated as follows. One may suppose that the translucency of consciousness varies in inverse proportion to the presence of the hyletic element. In states that contain more of unformed matter (which binds them all the more intimately to the bodily substratum) there is less of translucency; in states that contain less of hyletic matter there is more of the same. The socalled unconscious states are from this point of view the nearest approximations to the pure hyletic stuff, and in so far they are the least reflexive. As Ricoeur has so well argued, to speak of unconscious desires and hatreds is misleading in so far as it may lead one to imagine that there are in the unconscious fully formed desires or hatreds only lacking in the quality of consciousness.[93] While thus one should avoid what Ricoeur has called the Realism of the Unconscious, one should also avoid the other extreme view according to which the unconscious is the "mere impressional matter not yet brought to life by an intentional aim", the view namely that the unconscious is not yet intentional, not yet *of* ... On the latter view, defended by Ricoeur himself, the socalled unconscious desire, for example, is what the 'impressional matter' *would be* when it is made animated by an intention and thereby made conscious. This view leaves it unexplained why, if there is no unconscious intentionality and if it is the psycho-analyst who helps to transform the indeterminate matter into a desire or a hatred, then does the patient 'recall' from the depths of his memory and identify the disturbing element as a desire or a hatred which *was* there in his mind? Further, why should the disturbing matter be interpreted as a hatred and not as a suspicion, why as a love and not as a fondness? Is the intentional form which the psycho-analyst confers entirely arbitrary? We are therefore led to the hypothesis that the unconscious is already intentional, that it is already *of* ...; yet the intentional formation is vague, the mere pattern is laid down but not its definite meaning. The hyletic stuff predominates over the intentional meaning. The lack of transparency is accounted for by the presence of the hyletic stuff. But it is not entirely opaque, as no intentional act can be. For why otherwise does the patient 'recognise' and 'recall' it?

We may thus speak of degrees of intentionality: the bodily, the preconscious, the horizon intentionalities, the intentionalities of emotional and affective life and the intentionality of knowledge. Those intentionalities which are most transparent i.e. in whose case to have the act is also to be aware fully that one has it, are precisely those in whose case the hyletic stuff is almost absent. In knowledge in the strict sense the intention stands fulfilled, there is no content which obstructs the reference, no representation. The cognitive consciousness is the nearest approximation to the empty consciousness, for here the entire emphasis is on the object side: this is the element of truth in the modern theory that knowledge is not a mental state and that to describe one's knowledge is to describe the object one knows and not one's own mental state. For in knowledge consciousness is so fully absorbed in the object that the act has been emptied of all contents and approximates to the ideal of a pure consciousness that is nothing. And for that very reason we have here the highest degree of transparency: one cannot fail to be aware as and when one knows.

Our thesis then is this: intentionality is a necessary condition of reflexivity. But the more that is needed to constitute the necessary and sufficient condition is the relative absence of unformed hyletic stuff in consciousness.

This doctrine of degrees of intentionality and degrees of transparency of consciousness is consistent with Husserl's later doctrine that even transcendental self-experience is not entirely adequate.[94] He never accepted a total coincidence of consciousness with itself, just as he never accepted a total opacity in it. As Fink points out in his note to the *Krisis* book, the intentional analysis of phenomenology destroys the appearance of '*unmittelbarer Gegebenheit des Bewusstseins*'.[95]

University of Calcutta

REFERENCES

* Cf. The essay by A. Lingis in the present volume. (The editor.)

1 Special references may be made to: H. Spiegelberg, 'Der Begriff der Intentionalität in der Scholastik, bei Brentano und bei Husserl', *Philosophische Hefte* **5** (1936), 75–91; L. Landgrebe, *Phänomenologie und Metaphysik*, Schröder, Hamburg, 1949, pp. 59–69; A. Gurwitsch, 'On the Intentionality of Consciousness' in *Philosophical Essays in Memory of Edmund Husserl* (ed. by M. Farber, Harvard University Press, Cambridge, Mass., 1940; A. De Waelhens, 'L'idée phénoménologique d'intentionnalité', in *Husserl et la Pensée Moderne* (ed. by H. L. Van Breda and J. Taminiaux), Martinus Nijhoff, The Hague, 1959.

2 In fact, Chisholm regards this as the ontological part of the Brentano thesis. Cf. Chisholm's essay on 'Intentionality' in the *Encyclopedia of Philosophy* (ed. by P. Edwards), Colliers & Macmillan, New York, 1967.

3 F. Brentano, *The True and the Evident* (English translation by R. M. Chisholm *et al.*), Routledge & Kegan Paul, London, p. 78.

4 F. Brentano, *Psychologie vom empirischen Standpunkt*, Vol. II, Leipzig, 1911, pp. 133–138.

5 *Ibid.*

6 E. Husserl, *Logische Untersuchungen*, 4th ed., Max Niemeyer, Halle, 1928 (to be henceforth referred to as LU), II.1., pp. 370ff.

7 Ibid, p. 369.

8 Cf. E. Husserl, *Phänomenologische Psychologie*, *Husserliana*, Vol. IX, Martinus Nijhoff, The Hague, 1962 (to be henceforth referred to as P. Ps.), p. 268; *Erste Philosophie*, I, Martinus Nijhoff, The Hague, 1956 (to be henceforth referred to as EP), pp. 349; *Die Krisis der Europäischen Wissenschaften und die Transzendentale Phänomenologie*, Martinus Nijhoff, The Hague, 1954 (to be henceforth referred to as *Krisis*), p. 236.

9 P.Ps., pp. 314, pp. 260–266.

10 According to G. Berger, the category of *noema* is more fundamental than that of being or of non-being. Cf. G. Berger, *Le cogito dans la philosophie de* Husserl, Paris 1941, p. 54.

11 A. Gurwitsch, *Studies in Phenomenology and Psychology*, Northwestern University Press, 1966.

12 R. M. Chisholm, 'Sentences about Believing' in *Proceedings of the Aristotelian Society*, Vol. LVI, 1955–56, pp. 125–147; 'On some Psychological Concepts and the "Logic" of Intentionality' in *Intentionality, Minds, and Perception* (ed. by H. N. Castañeda), Wayne State University, 1966, pp. 11–35; also his Encyclopedia article.

13 L. Landgrebe, *loc. cit.*

14 H. Spiegelberg, *loc. cit.*, p. 86.

[15] Contrast Quinton, 'Mind and Matter' in *Brain and Mind, Modern Concepts of the Nature of Mind* (ed. by J. R. Smythies), Humanities Press, New York, 1965. Quinton argues (p. 224) that the Brentano thesis rules out behaviourism. It surely does, if intentionality is regarded as was done by Brentano himself as a criterion of the mental, and if behaviour is understood in a mechanistic way, not otherwise.
[16] A. Diemer, *Edmund Husserl, Versuch einer systematischen Darstellung seiner Phänomenologie*, Meisenheim, 1956.
[17] Cf. J. N. Mohanty, *Edmund Husserl's Theory of Meaning*, Martinus Nijhoff, The Hague, 1964, pp. 80–86.
[18] Cf. P. Ricœur, *Freedom and Nature. The Voluntary and the Involuntary* (transl. by E. V. Kohak), Northwestern University Press, 1966, pp. 205ff.
[19] Cf. P. Ricœur, *Husserl, An Analysis of his Phenomenology* (transl. by E. G. Ballard and L. E. Embree), Northwestern University Press, 1967, p. 192.
[20] T. W. Adorno, *Zur Metakritik der Erkenntnistheorie, Studien über Husserl und die phänomenologische Antinomien*, Kohlhammer, Stuttgart, 1956, p. 171.
[21] E. Husserl, *Ideas* I (references will be made to the Boyce Gibson translation in the paperback edition which will be henceforth referred to as *Ideas* I), p. 250.
[22] A. Gurwitsch, *Studies*, pp. 138–139.
[23] *Ideas* I, p. 267.
[24] For elaboration of this point, see J. N. Mohanty, 'Notas a las lecciones de Husserl sobre la conciencia del tiempo', Dianoia, Anuario de Filosofia, Mexico, 1968, pp. 82–95.
[25] R. Sokolowski, *The Formation of Husserl's Concept of Constitution*, Martinus Nijhoff, The Hague, 1964, pp. 92–93, 100, 114–115.
[26] *Ideas* I, p. 226.
[27] Gaston Berger, bearing witness to Husserl's later thought in this matter, refers to "how much the word 'construction' could irritate Husserl" (Gurwitsch, *Studies*, p. 160). For an account of the different stages of Husserl's attitude towards the notion of productive synthesis, see Iso Kern, *Husserl und Kant*, Martinus Nijhoff, The Hague, 1964, pp. 249ff.
[28] P. Ricoeur, *Husserl*, p. 41.
[29] E. Husserl, *Ideen*, I (Husserliana edition), p. 100. (This edition will be referred to as *Ideen* I.)
[30] *Ideen* I, § 82.
[31] *Ideen* I, § 95.
[32] *Ideen* I, § 117.
[33] *Ideen* II (Husserliana, Vol. IV), Martinus Nijhoff, The Hague, 1952, § 4.
[34] E. Husserl, *Analysen zur Passiven Synthesis* (1918–1926), Martinus Nijhoff, The Hague, p. 196.
[35] *Cartesian Meditations*, translated by D. Cairns, Martinus Nijhoff, The Hague, 1960, § 38 (to be henceforth referred to as CM).
[36] P.Ps., pp. 98–99.
[37] *Formale und Transzendentale Logik* (to be henceforth referred to as FuTL), Max Niemeyer, Halle. 1929, p. 184.
[38] R. Sokolowski, *loc. cit.*, p. 172.
[39] G. Funke, *Zur transzendentalen Phänomenologie*, Bonn 1957, pp. 12–13; CM § 32.
[40] Thus T. Seebohm, '*Die Bedingungen der Möglichkeit der Transzendental-Philosophie*, Bouvier, Bonn, 1962: "... ist die genetische Intentionalanalyse auf die Zusammenhänge dieser Typen in der Einheit des Bewußtseinsstromes selbst gerichtet." This surely is not *eo ipso* a historically oriented programme.

41 For example, in P.Ps., p. 431.
42 CM, § 19.
43 CM, § 64.
44 CM, p. 47.
45 *Krisis*, p. 240.
46 EP, II, pp. 318–319.
47 P.Ps., p. 428.
48 FuTL, p. 208; § 94.
49 G. Brand, *Welt, Ich und Zeit, nach unveröffentlichen Manuskripten E. Husserls*, Martinus Nijhoff, The Hague, 1955, p. 23.
50 E. Fink, *Studien zur Phänomenologie*, Martinus Nijhoff, The Hague, 1966, p. 219.
51 M. Merleau-Ponty, *Phenomenology of Perception*, London, 1962, Introduction.
52 J. J. Kockelmans, *Phenomenology and Physical Science. An Introduction to the Philosophy of Physical Science*, Duquesne University Press, 1966, pp. 53ff.
53 J.-P. Sartre, *The Transcendence of the Ego. An Existentialist Theory of Consciousness*, translated by F. Williams and R. Kirkpatrick, Noonday, New York, 1957.
54 J.-P. Sartre, *Being and Nothingness. An Essay on Phenomenological Ontology*, Translated by H. E. Barnes (to be henceforth referred to as BN), Methuen, London, 1957, p. lxi.
55 BN, p. 4.
56 BN, p. lxiii.
57 BN, p. 109.
58 BN, p. lix.
59 L. Landgrebe, 'Prinzipien der Lehre vom Empfinden' in *Zeitschrift für philosophische Forschung*, **8** (1954) 193–209.
60 H. U. Asemissen, *Strukturanalytische Probleme der Wahrnehmung in der Phänomenologie Husserls, Kant-Studien Ergänzungshefte* **73**, 26–34, Kölner Universitätsverlag, 1957.
61 M. Merleau-Ponty, *loc. cit.*, 243. Also see the same work, pp. 213, 267, 405.
62 This is so according to Diemer.
63 BN, p. lx.
64 BN, p. lxi.
65 T. Langan, *Merleau-Ponty's Critique of Reason*, Yale University Press, 1966, p. 23f.
66 M. Merleau-Ponty, *loc. cit.*, p. xvii.
67 Cf. *ibid.*, pp. 124–125.
68 Cf. *ibid.*, pp. 57, 58–59. Also, Merleau-Ponty, M., *Signs* (transl. by McCleary), Northwestern University Press, 1964, p. 173.
69 M. Merleau-Ponty, *Phenomenology of Perception*, p. 137.
70 M. Heidegger, *Vom Wesen des Grundes*, Klostermann, Frankfurt, 1949, pp. 15, 44.
71 BN, p. 85.
72 P. Ricœur, *Freedom and Nature*, 61.
73 This is the argument of Natanson in his 'The Empirical and Transcendental Ego' in *For Roman Ingarden. Nine Essays in Phenomenology* (ed. by A. T. Tymieniecka), Martinus Nijhoff, The Hague, 1959.
74 *Ideen* II, § 28; but the doctrine of the autogenesis of the ego is surely suggested in the same work, p. 102.
75 *Ibid*, p. 310f.
76 P.Ps., pp. 207–208.
77 CM, § 32. On 'habitualities', see P. Ricœur, *Husserl*, pp. 54–55.
78 G. Funke, *loc. cit.*, pp. 22–23.

[79] Seebohm, e.g., holds that the concept of *hyle* may be non-sensualistically interpreted within the framework of the intentionality thesis. See *loc. cit.* p. 97 fn. 18.
[80] P. Ricœur, *Freedom and Nature*, pp. 373–409.
[81] *Ideas* I, § 53.
[82] *Ibid.*, § 53.
[83] Section 2, Chapter 3.
[84] *Ideen* II, pp. 56, 158f.
[85] *Ibid.*, p. 151.
[86] *Ibid.*, p. 151f.
[87] *Ibid.*, p. 153.
[88] *Ideen* III, p. 118.
[89] *Ibid.*, p. 124. From this point of view, Ricœur's statement (in his *Husserl*, p. 61) that the distinction between existence and objectivity is not known to Husserl may only be conditionally accepted.
[90] CM, § 28.
[91] P.Ps., p. 197.
[92] *Ideen* I, § 38.
[93] P. Ricœur, *Freedom and Nature*, pp. 387–389.
[94] CM, esp. § 9.
[95] *Krisis*, p. 474.

CORNELIS A. VAN PEURSEN

THE CONCEPT OF THE BODY IN TRANSCENDENTAL PHENOMENOLOGY AND IN MODERN BIOLOGY

I. HUSSERL'S PHENOMENOLOGY OF THE BODY

Rich material as to the human behavior and experience related to the body has been brought forward by psychology and psychopathology during the last decades. These new conceptions are generally called 'phenomenological', and many of them have, indeed been influenced by phenomenological philosophers, like the earlier Sartre and like Merleau-Ponty. Others, like P. Schilder and Erwin Strauss have developed very penetrating analyses along their own lines and even before the aforementioned thinkers had evolved their ideas. The influence of Husserl on the greater part of this current is undeniable, but is nevertheless restricted to the direct or indirect influence of his general phenomenological approach. Many of his thoughts on the body are worked out only in his later works, and where they are found already in earlier writings, such as in the second volume of his *Ideas*, they remained unpublished. *Ideas, II* was published only in 1952, many years after Husserl's death. It is useful, therefore, to analyse now Husserl's analyses of the body and their importance for Husserl's concept of a transcendental phenomenology, and to draw some lines of comparison with contemporary results of biological and physiological research.

It must be clear from the beginning that Husserl's analyses of the body are situated in quite another dimension than those of contemporary phenomenological psychology. The latter tries to clarify the meaning of empirical attitudes and patterns of behavior of man by taking account of differences in the experience of the own body. It is, for instance, quite normal behavior when somebody tries to find an object, even if this object is the part of the body of another; but when he tries to find one of his own limbs his behavior is no longer normal, and the very experience of his own self must be disturbed. Such analyses, to be found in the works of various psychologists, psychiatrists, physiologists, and also of philosophers like Sartre, imply, of course, certain philosophical and even phe-

Tymieniecka (ed.), Analecta Husserliana, Vol. I, 133–151. *All Rights Reserved*
Copyright © 1970 *by D. Reidel Publishing Company, Dordrecht-Holland*

nomenological categories, like those of 'subjectivity' and 'subject-object correlation', 'horizon' (W. James: "fringe"). Their aim remains, however, the research into empirical attitudes and processes. And here lies the difference with Husserl's phenomenology, which during all the varied stages of its evolution, has had only one theme: the clarification of the purely logical structure of phenomena.

It is not the intention of this article to give even a succinct summary of Husserl's analyses of the human body; that would require a whole book, and has been done already by competent authors.[1] It is only useful to illustrate the previous thesis about the difference between Husserl's phenomenology of the body and that of phenomenological psychology, in giving a few examples of Husserl's analyses throughout his works. This might demonstrate the continuity of his thought on this point, at the same time stressing the prevailing logical character of his analysis.

Many contemporary philosophers are united in their rejection of the Cartesian separation of body and mind. Analytic philosophy points to the misleading character of language when the relationship between the mental and the physical is rendered by expressions that refer to the mind as if it were a thing to be found in another thing (the body). A clarification of the meaning of language has to start with the elimination of this source of philosophical bewilderment. We must refuse to turn activities, like thinking, into substantives and finally into substances (*res cogitans*). Authors like Wittgenstein and Ryle mention the misleading image of the mind as a thing to be found in a box (body).[2]

Husserl's phenomenology, from the beginning a logical investigation replacing any psychologistic reduction of logical laws to empirical rules of thinking, implies a similar rejection. He uses the same image already in his *Logische Untersuchungen* in order to reject it; but he starts at an earlier level, pointing to the idea that the representations are thought of as being objects within a box (the consciousness). He replaces the image of object-in-a-box by a dynamic one, replacing thus the substantive by a verb: the objects of which we are aware are not to be found in our consciousness like objects in a box, but they are being constituted by a variety of acts of intentionality. A new program of logical investigation is disclosed in this way: the differences between various representations (a perceived house, a mathematical proposition, a memory, etc.) cannot be discovered when one investigates merely the differences as different contents of representation, but only when one investigates the different ways in which consciousness is being directed towards objects.[3]

Husserl makes it clear, in his further investigations, that the analysis of the directedness of consciousness does not concern the empirical state or processes of a given human being, but deals with the rules that delineate the various meanings of acts ('perception', 'memory', 'mathematical understanding'). Not the empirically given states and processes of consciousness are decisive for the acknowledgment of a certain act as 'perception', 'memory' etc., but rather the logically definable rules determine whether or not we have the right to call a given act 'perception' etc. It is like the game of chess: the empirical behavior does not constitute the classification; certain patterns of human behavior are classified as 'playing chess' only when they are in accordance with certain rules. Husserl's analysis of the directedness of consciousness is not a psychological but a logical one.

When consciousness is redefined in terms of activity, regulated by logical rules, the body itself also needs a more logical approach. In his writings since 1912, Husserl gives various examples of such redefinitions. He writes that the tactile perception is not localized in the touching finger and that, likewise, the process of thinking is not localized in the head. Of course, he recognizes that empirical experiences of certain physiological tensions or tactile pressions can be localized in the head or finger, but these are only the empirical-physiological components of the intentionality by which the apperception of the touched object or of the content of a thought takes place. This means that consciousness (of, for example, a touched object or of a taught mathematical proposition) is not at all something inside of the body, nor even a special layer (*Schicht*) within the total corporeal existence. The intentional experiences are united with the body through the empirical-physiological ('hyletic') components, but are as such nothing that can be formulated in the ways that things are (*nicht-dinglich*).[4]

The human body can be taken as a material object among material objects, and is then subject to laws of causality. The human body can also be taken as the center of sense-experience. The relationship between physiologically localizable processes (such as of tactile character) cannot, in this case, be taken as causes of non-physiological experiences. The relationship here is the one of implication ('if–then'): I experience sensation when my hand is being touched, stung, etc. This means that there exists in nature a material object, which is not merely a material object but also a body. 'Body' is a material object that is at the same time the field of localization of sensations and feelings, and as such a fundamental part of the given self (*Ichgegebenheit*).[5]

In such ways Husserl analyses the bodiliness of man, showing how in the empirical data of processes intentional activity can be distinguished according to certain rules. Many more features are analyzed along these lines and Husserl offers convincing descriptions, such as of the orientation in space, with 'here' and 'there', the body as necessary point of zero within coordinates of orientation, the differences between moving an object and moving one's own body, etc. The dimension of subjectivity becomes apparent in all these analyses. In the same way that the more logical relationship of implication cannot be identified with mere physical (empirical) relationship of causality, the subjective dimensions of orientation etc. cannot be put on a par with the objectively given phenomena of the surrounding world. The direct perception of other human beings is, however, one of human bodies as centers of fields of perception and action. It is in this way that transcendental, that is, logically analyzable rules of subjectivity, are visible in the empirically given world.

Other important aspects of this transcendental (that is, not empirically reducible to the meaning of the constituting empirical data) dimension are related to the analysis of the body. The body is present at what is taking place around man, and so the notion of the 'now', an important theme of the whole development of Husserl's phenomenology, is involved here.[6] The body is also the place of the kinaesthetic experiences: two aspects are here united, the objective movement in space and the kinaesthetic movement of the subjective awareness that I am moving. In his *Ideas*, II and in wider, more dynamic contexts in later writings, Husserl points to this aspect of a dimension of subjectivity (*Ich-Seite*) as the field of research for a phenomenology of bodiliness (*Phänomenologie der Leiblichkeit*). This has to result in a phenomenological clarification of subjectivity as it becomes apparent in empirical nature (context of the human body).[7]

Empirical data and transcendental-logical rules are interrelated. Nevertheless, it is exactly this interrelationship that constitutes one of the most serious problems of Husserl's phenomenology. Husserl speaks often about a radical separation between a more empirical or psychological analysis of subjectivity and a transcendental one. Psychology situates its object within the limitations of a really given world. The mistake of any philosophy trying to prove the existence of the outside world *via* a causal argument is that it starts from a pure, in itself existing Ego, as if this were a part of the natural world. The above-mentioned relation of, for instance, physiological processes to the logical dimension of subjectivity is being described now as a process of correlation between constituting consciousness and constituted world, a correlation to be found in the perspective of transcendental subjectivity.[8]

This transcendental Ego or subjectivity is, in a way, the logical consequence of the first steps of the *Logical Investigations*. Everything that is represented, including the own psychophysical existence, is being liberated from empirical contingency as it is analyzed according to the rules visible from a transcendental point of view. The various forms of reduction are disclosing this logical field as a process of constitution, through active and passive syntheses, a constitution to be transcendentally qualified as the achievement (*Leistung*) of the Ego. Representations are not things in a box and so the whole structure of consciousness – as to its logical aspect of consciousness-of-something, of intentionality, not as to its physiological and psychological components – has to be taken as formal rules, not as material (hyletic) data. The validity of the grammar – already in his *Logical Investigations* Husserl refers to a logical grammar[9] – cannot be deduced from the empirical text (Husserl: matter of fact), but the empirical text can only be read when one knows the rules of the grammar.

The difficulties arise when one asks about the status of the transcendental Ego. It has no metaphysical status and it must not be hypostatized, as Husserl often says. As early as in 1905, in his lectures on time-consciousness, Husserl states that the 'now' as the source of time cannot be described in structures derived from the process of temporalization and objectification. It cannot be expressed in words at all; "for all this names are lacking".[10] When Husserl relates, in later works, the 'now' to the position of the transcendental subjectivity, he continues in the same line by calling this Ego 'anonymous'. His later analyses of the life-world can be considered to be the investigation of the logically analyzable process by which the anonymous Ego acquires names; that is, becomes constituted as man in the world. The Ego has to be brought to expression (*zur Auslegung*, *zur Aussprache*). It is as such not declinable, and becomes declinable only by a special constitutive achievement (*Leistung*).[11] It is not possible to elaborate this theme further here and to describe in more detail this transcendental constitution. Many authors have done this, pointing to the difficulties related to such an analysis of the life-world as field of constitution of a 'lonely', transcendental Ego.[12]

It can be stated, in conclusion, that Husserl's phenomenology of the body is no aim in itself. He wants to clarify, through the medium of the orientation in space and time of the human body, the logical lines converging in the dimension of subjectivity. This dimension is more than merely psychological; it is logical in the sense that it constitutes the presupposition for any other analysis of these phenomena, by any scientific discipline, just as a grammar is presupposed for the understanding of spoken or written words. The problem is how such a transcendental dimension can be indicated as in itself 'not declinable', 'pure', 'absolute', 'anonymous', and how, on the other hand, its declination (that is, its empirical realization in human names, history and life-world) is necessary from a transcendental point of view.

The following pages will try to show the importance of this dilemma that has never been solved by Husserl. It will be useful to point to similar problems in recent analyses, partly philosophical, partly scientific, that are more or less analogous to Husserl's dilemma – he speaks himself often of a 'paradox'. The direction to follow for a possible solution will then become visible also.

Two final remarks, underlining this paradox, have to be added. The first is that Husserl needs the concrete and given phenomena, including the contingent world (nature) and the limited human body, in order to make manifest the logical rules. He needs the description of the contingent in order to show the non-contingent. The 'transcendence in the immanence' is a frequently used expression that indicates the result of the various reductions. All this implies a positive and logical outcome of Husserl's analysis of the body: in the contingent, human existence becomes visible, through the bodily subjectivity and orientation (transcending the finiteness), generally valid rules, and pure phenomenological processes. This 'surplus' of transcendence in immanence requires, for an explanation, the activity (achievement) of the transcendental Ego. Thus, the empirical world and the pre-theoretical horizon of the life-world, including the direct given experiences of the human body, are necessary; and this by their contingency, to make manifest the necessity of the transcendental Ego. Husserl, states in various ways that truth and evidence are never given as such, but necessarily in perspectives of anticipation and 'presumption': the limitations are, in that way, essential laws (*Wesensgesetz*).[13]

The second remark is that it is for Husserl methodologically impossible to go the other way round and to demonstrate that there is a necessary implication of the empirical, including the human bodiliness, in the transcendental (the Ego). This point has been discussed already, so it is sufficient to refer to only one theme, not mentioned as yet, the 're-iterations': empirical man, bodily center of a field of orientation, can refer back to his own subjectivity, disclosing in this way, *ad infinitum*, possibilities of referring to the own self ('me') from the standpoint of a new level of subjectivity ('I'). I see a house; I know that I see a house; I know that I know ... etc. There are, however, no iterative acts that lead really into the opposite direction, in that of the endless necessity of evaluations. Or, in other words, the reflexion is always – even as reflexion on the transcendence in the immanence – self-reflexion in a transcendental direction; it never has the character of a response and it lacks, in this sense, a final ethical dimension. The difference with other phenomenologists like Thévenaz, Levinas and Ricœur is here striking. Methodologically, this implies that the 'declination' of the 'lonely' Ego cannot receive any explanation. Organization, also in methodological sense, implies control as well as adaptation; the latter aspect is lacking in Husserl's analyses, and so he comes to statements without further possible foundation, ('consciousness is not conceivable without any organization'); even to metaphysical assertions that this Ego is "immortal' and 'eternal'.[14]

II. A LINGUISTIC ANALYSIS OF THE BODY

Only a few illustrations will be given in order to point to some analogous thoughts in the field of analytic philosophy. A full discussion of this topic would be interesting, but cannot be undertaken here within the limits of the clarification of Husserl's position. The aim is only to stress the previous conclusion about the dilemma of a mere transcendental analysis of the human body. Linguistic analysis starts from the opposite direction, and here the difficulty is more the other aspect: how to demonstrate that non-logical uses of language, especially those of ordinary language connected with the human body, show nevertheless certain regularities and rules? Our next paragraph, giving a concise review of the results of scientific investigations, especially those concerning the CNS (central nervous system), has to give concrete material enabling us to reconcile both aspects and to solve the dilemma.

Ludwig Wittgenstein's work is the main source for these analyses. His early work, the *Tractatus*, contains some remarks on the Ego that show a striking parallel with Husserl's explanations, and are perhaps in some way influenced by them. Their perspective, however, is quite different; Husserl was concerned with the *a priori* structure of the Ego, while Wittgenstein denies any *a priori* order of things. Husserl finds many *a priori* structures, and therefore speaks about the Ego as such that is, nevertheless, not expressible without a name. Wittgenstein concludes his work with the well-known statement that what we cannot speak about we must consign to silence; there is only logical necessity, and 'logic' is for him much narrower than for Husserl, as it is confined to tautological rules to be formulated by symbolic logic.

Considering these differences, the resemblance of some statements of Wittgenstein's to those of Husserl remains striking. A longer quotation may demonstrate how Wittgenstein was facing, from a different angle, the same problem as Husserl.

> The thinking, presenting subject; there is no such thing. If I wrote a book 'The world as I found it', I should also have therein to report on my body and say which members obey my will and which do not, etc. This then would be a method of isolating the subject or rather of showing that in an important sense there is no subject: that is to say, of it alone in this book mention could not be made. The subject does not belong to the world but it is a limit of the world. Where in the world is a metaphysical subject to be noted? ... There is therefore really a sense in which in philosophy we can talk of a non-psychological I. The I occurs in philosophy through the fact that the 'world is my world'. The philosophical I is not the man, not the human body or the human soul of which psychology treats, but the metaphysical subject, the limit – not a part of the world.[15]

In Wittgenstein's later works, there is a clear shift in his analyses of the human body. Also in his study the dimension of subjectivity does not remain inexpressible, but characterizes the bodily position of man. In his *Philosophische Bemerkungen* of 1930 and in his *Zettel* (between 1929 and 1948), Wittgenstein refers often to the kinaesthetic sensations.

> It is queer. My lower arm is now resting horizontally, and I should like to say that I feel it; not, however, as if I had a feeling that always accompanies this position (as one would feel ischaemia or engorgement) – but as if the 'body-feeling' of the arm were arranged or disposed horizontally ...

He qualifies the space, in which the body moves itself and orientates itself, as a space of perception as well as of muscular sensation. This space of bodily orientation is not symmetrical as is physical space. There is a secondary type of language in which the physical space is called 'objective' and the space of perception 'subjective', but this 'physical space' is a construction, having as a basis the space of perception.[16] Such an analysis, to be found in *Philosophische Bemerkungen* runs parallel to Husserl's later analyses of geometrical space as being derived from, and constructed on the basis of, the space of the lifeworld.

This recognition of the subjective dimension of bodily orientation leads Wittgenstein to a kind of phenomenological (he uses the term sometimes himself, although in a somewhat different sense) distinction between what can be physically objectified (a machine) and what cannot (the body).

> How does it come about that I see the tree standing up straight even if I incline my head to one side, and so the retinal image is that of an obliquely standing tree? ... But must there be a physiological explanation here? Why don't we just leave explaining alone? – But you would never talk like that, if you were examining the behaviour of a machine! – Well, who says that a living creature, an animal body, is a machine in this sense? –[17]

Wittgenstein's analyses of the body do form, of course, only a part of the wider range of his analysis of language, in which other dimensions than the purely objective and logical one also become apparent. His investigation of bodily orientation takes place through the medium of language, where different language-games, each with its own rules, are distinguished.

The dimension of subjectivity constitutes a new type of language-game that can be applied to the analysis of the human body. The change in position compared with the *Tractatus* is, in this respect, that now in a book, entitled 'The world as I found it', there would be place for the subject. There are two types of propositions, states Wittgenstein in his *The Blue and Brown Books* (of 1933–1935), one describing facts in the material, external world, and the other describing personal experiences. Both types are mixed up – and here we get into grammatical troubles – in questions like 'Is it possible for a machine to think?'. Wittgenstein clarifies this point further by distinguishing two linguistic dimensions related to two aspects of the body. He speaks of 'the geometrical eye' and 'the physical eye'. The first is involved when I bring my hand to my eye, without seeing my eye as a physical object; it is 'the kinaesthetic experience of raising my hand to my eye'. The second is involved when I look into a mirror and see my finger nearing my eye. Both are dimensions, not two different objects; much grammatical bewilderment is eliminated (such as that about mind and matter) when these dimensions are distinguished. They are directly related to the word 'I' – and here the analyses come close to those of Husserl.

> Now the idea that the real I lives in my body is connected with the peculiar grammar of the word "I", and the misunderstandings this grammar is liable to give rise to. There are two different cases in the use of the word "I" (or "my") which I might call "the use as object" and "the use as subject". Examples of the first kind of use are these: "My arm is broken", "I have grown six inches" ...Examples of the second kind are: "I see so-and-so" ... "I try to lift my arm" ...[18]

The difference with Husserl's phenomenology of the body is that the recognition of a dimension of subjectivity, which plays an important rôle in Wittgenstein's analysis, does not lead to a transcendental dimension. It never becomes the zone where the other achievements in orientation and in language are constituted. Wittgenstein speaks in his later works (*Philosophical Investigations* of 1937–1949, and *Remarks on the Foundations of Mathematics* of 1937–1944) of the more logically scheduled languages of mathematics and physics as constructions related, in some ways, to the varied dimensions of daily language, which, in its turn, has to be seen in connection with the human 'forms of life'. This theme is, evidently, similar to Husserl's life-world. But the constructions of more logical and ideal languages are, in Husserl's late phenomenology, the expressions of the achievement of the transcendental Ego. They go back, for Husserl, to a transcendental logic. The forms of life and the rules of their language-games are, for Wittgenstein, not reducible further to any logical or transcendental process.

This means that the dimension of subjectivity, the 'I', can be used for the analysis of the body, but not for any analysis of the sciences. The word 'I' can serve to distinguish me from other people, as when I say 'I have pain', but it does not indicate a certain person in an objective way. "'I' is not the name of a person, nor 'here' of a place, and 'this' is not a name; but they are connected with names. Names are explained by means of them. It is also true that it is characteristic of physics not to use these words."[19]

Husserl tries, in a sense, to incorporate the objective, physical dimension of reality into the subjective, that is, transcendental one. Bodily orientation can illustrate this. Wittgenstein maintains, at least in general, a sharp distinction between a physical and a more subjective language, recognizing at a few places that the more objective language is a construction. Other analytic philosophers elaborate his views and come closer to a phenomenological position. Stuart Hampshire stresses the intentionality of the body. The human observer must be thought of as a persisting body among others, he has to be situated bodily in order to be able to point to things. Thus knowledge, of sciences as well as of daily experience, is impossible as an act of an unsituated consciousness; it requires a bodily position and intentionality.[20] All these lines are converging into the question of how the factual, contingent position of the body can demonstrate such regularities as are not those of mere objectivity, but rather which render possible any orientation presupposed by the objective constructs of scientific knowledge. Husserl answers that this has to be interpreted as an achievement referring to the operations of a transcendental Ego. But for what reasons are these logical operations realized in the empirical body and the empirical world, why is there any declension of the Ego? The answer of Wittgenstein is that the various patterns of human behavior, including the kinaesthetic ones, are related to different ways of using words, such that different grammars must be used. But what are the relations between physical language, where words like 'I' and 'here' are not used, and the subjective language of the 'geometrical eye' that nevertheless constitutes the ground for the constructions of more rigidly scheduled types of language?

III. THE HUMAN BODY AS THE SYNTHESIS OF RULES AND EMPIRICAL DATA

Husserl's analysis of human bodiliness is more than a psychology of the experienced body; it is an effort to disclose valid rules in empirical data, a scheme of orientation surpassing the limited situation, a transcendence in immanence, a transcendental dimension and achievement within the naturally given world. An analysis starting from the other end with the empirical patterns of language, as has been given by Wittgenstein, can in many respects underline the necessity of a more logical or structural investigation of such a directly and empirically given 'object' as the human body. It is useful to derive some more arguments from another field, with which phenomenology and analytic philosophy have for various reasons had little contact: the biological sciences. These demonstrate, in recent development, the same necessity for a more transcendental reflexion, as Husserl carried out, giving at the same time some hints for the solution of the dilemma of Husserl's transcendental idealism. The trends in those sciences, as related to the investigation of the structure of the body, will be discussed here only very briefly as they serve only as illustration for the question posed at the end of the last paragraph.[21]

The body of an animal being is genetically defined by the structure of the deoxyribonucleic acid (DNA), discovered in 1944 and structurally analyzed in 1953 by Watson and Crick. This chromosomal DNA is involved in protein synthesis and determines, through the constitution of albumins, all hereditary properties of a living being. There are 20 amino acids out of which proteins can be synthesized, to be compared with an alphabet through which a text is composed. This composition is commanded by DNA, which plays thus the role of a matrix. The DNA-code itself is composed of only 4 elements (*A*denine, *C*ytosine, *G*uanine, *T*hymine), millions of molecules of which are attached to each other in long threads, twisted into a helix so that A opposes always T, and C opposes G. Such a matrix (of which one thread might for example read GAGGTCACCTGGAT... etc.) controls the constitution of human bodiliness (color of eyes, constitution of the brain, temperament, etc. etc.).

The sequence of chemical elements in nature results from processes that obey the laws of probability. These laws are the mathematical expression of the physical law that processes in nature will tend to level the differences among energetic constellations or, in other words, will augment 'disorder' (the elimination of specific and exceptional structures). The sequence of chemical elements in DNA is, however, not contingent, not obeying the laws of probability. The result of the functioning of DNA in genetic processes becomes apparent in the specific structures of the body of a living being. DNA can, therefore, be considered as a code that guarantees the transport of information: it is not a contingent and haphazard sequence, but a series of biochemical commands, the blueprint of the complex organization of the body. The body is the realization and reification of this genetic code.

It is possible to give a mathematical formulation of probability. It is also possible to calculate mathematically how far a given sequence deviates from the most fortuitous order. The measure of probability is called 'entropy', and indicates the measure of accordance with the second law of thermodynamics (the tendency of differences in energy levels to be eliminated); the measure of deviation is called 'negentropy' or 'information' and indicates the measure of order. This information-content – in formal-mathematical, not in material-semantic sense – is expressed in binary digits, 'bits'. Information-theory has been developed in the fields of electronics and mathematics by men like Shannon, McKay, and Brillouin, and has been applied to the field of biology by biochemists and biologists like Perutz, Quastler, and Raven. The evolution of living beings shows a continuous increase in bits, running parallel with specialization and more specific structures of bodily organization. The human body especially is defined by hereditary information of unimaginable size (estimated by Dancoff and Quastler at 10^{15} bits). Evolution can be seen, under this perspective, as the rise of order and of rule out of lower levels or organization. Have mathematically analysable rules their source in mere empirical contingencies? How is such an emergence to be explained? Or is the question misleadingly formulated?

The very specific organization of man culminates in the constitution of his central nervous system (CNS). This CNS is itself required by the blueprint of the human DNA. It is the source of another type of information: not that genetically transported by the chromosomes, but the information to be gained during the life-time, acquired information. Every type of knowledge can, in this respect, be considered as order drawn from the surrounding world by the human brain. This order results in the organization of behavior and in the control of the surrounding world by man. This human behavior has a much wider range than animal behavior, as it can utilize symbols and need not, therefore, respond only to directly given stimuli, but can postpone responses, plan actions, acquire new types of response-patterns by learning-processes. All this means that the main function of the human CNS is not merely that of acquiring information. It goes much further in that it develops patterns of organization and rules for possible behavior.[22]

Three philosophically important implications of all this have to be mentioned. The first is that the human body is not a static datum, but a dynamic field of regulated behavior. The body has to be seen as agent, as center of activities – in more phenomenological terms, it must have an intentional structure. This point is, of course, equally applicable to the animal body. A second implication gives the first point, however, a specific turn for the human body: the bodily behavior is regulated by symbolic rules. The human CNS, with its evoluted cortex, has at its disposal a kind of switching circuitry characterized by logical rules. In 1938, Shannon applied truth-functional logic to the operations of switching circuits, and in 1943 McCulloch and Pitts interpreted neuronal behavior analogously as two-state switching devices. This demonstrates how the human CNS, in distinction to the CNS of other higher primates, can organize the human behavior according to maps of knowledge structured by rules of logical consistency. The human body can structure symbolically his surrounding world, gaining in this way much wider control over this world.

A third implication results from the previous ones. The human CNS, and this means also the total bodily behavior, has a specific centralized organization. Many authors point to the fact that the CNS of higher primates has various centers in its cortex for the coordination of operations, but that these centers are integrated into one in the human CNS. This means, writes biologist W. H. Thorpe, that there exists a plurality of 'spirits' in the animal, a plurality that becomes gradually integrated into the human spirit.[23] This hierarchical structure of the brain renders possible the human capacity for solving, along the consistent rules of knowledge, the problems presented to him by the external world. "We start by observing a living body with a primordial center of individuality, and are led on by a continuous progression to a situation in which we face a subject committing himself deliberately to the solution of an external problem", writes M. Polanyi.[24]

It is clear that this last point relates closely to, in the analysis of Husserl, the transcendental Ego as the third dimension of the human body. Biologists and other authors studying this field often refer, at any rate, to the dimension of subjectivity, as it becomes visible in the operations of the human CNS and thus in the whole pattern of human bodily behavior. There are two aspects, a more physiological and a more logical one. The first aspect is that the CNS is mirroring in a way all that which occurs in the human body – the physiological expression of human self-consciousness and of bodily self-orientation. It is well known how the motor-cortex is an area in which the various movements of limbs etc. can be functionally mapped out. It is as if here one finds the key-board of the motoric functions of the whole body. But, as Ruyer formulates it: remarkably enough, the player of the keyboard cannot be found; the cortex does not mirror itself.[25] A kind of unseizable dimension of subjectivity must be accepted.

This dimension of subjectivity of the body can also be viewed from a different, more logical, angle. The many, often highly technical, discussions cannot be summarized within the scope of this article, and they are therefore mentioned only in passing. They are related to the theorem of K. Gödel stating that in order to prove the consistency of a formal system one cannot restrict oneself to the media of that system but has to go outside the system; keeping a system formally closed implies the impossibility of proving its consistency. The mathematician Ladrière is of the opinion that Gödel's Theorem is the consequence of a self-referential structure of any formal system. It is through this self-referential aspect of logical thought that a transcending of the limits of a formal system takes place.[26]

It is clear that this limitation of a system of formal rules can be related to the investigation of the operations of the brain. At this moment, Husserl's thesis of the transcendence in the immanence as an achievement of the transcendental Ego finds a direct illustration in modern neurophysiological investigations. There are many authors making such a direct connection between Gödel's Theorem about formal systems and the way in which human consciousness and self-awareness functions. "The brain that tries to understand, is itself part of the world that it is trying to understand..." writes P. W. Bridgman. Also other authors stress this unique capacity of human consciousness to stand over and against the surrounding reality, even over and against his own existence and thinking. This is not any more a theme for philosophical speculation, but rather a necessary dimension in the interpretation of empirical processes such as human bodily orientation, symbolic behavior, and language.[27]

All this is still the material, derived from developments in biological sciences, which we need for an answer to the dilemma raised by Husserl's phenomenology of the body. In order to use this material for a conclusion a more philosophical interpretation of the aforementioned data and opinions is necessary. Husserl compares the logical rules of thought with the arithmetic rules controlling the operations of a calculating machine. Mechanically the machine and its workings have to be explained by physical laws, but the arithmetical operations cannot be explained by these physical laws. Husserl states that, in the same way, the rules of our thinking cannot be explained by the empirical laws of our thinking machine; logical rules cannot be clarified on the basis of empirical (physical, psychological) laws.[28] This distinction, used by Husserl in 1900 as argument against any psychologistic reduction of the rules of logic, is still valuable. One cannot, for the same reasons, explain logical and scientific rules by referring merely to the empirical machinery of the brain.

Neglecting for a moment the quite different approaches of Husserl, of Wittgenstein, and of the biologists and physiologists, a convergence of their opinions can be indicated: the empirical material concerned with the human body needs an interpretation which acknowledges the involvement of the subject in the objective explanation of bodily organization and symbolic behavior. The more transcendental approach of Husserl on the one hand and the logical investigations of brain-processes as well as the limitations of formal systems on the other constitute two different wings of the same effort to clarify empirical phenomena by a logically structured interpretation. It is only within such a more logical – in the wide sense of a methodological – perspective that the meaning and the ways of operation of the empirical processes can be disclosed. It is clear, for example, that only within the dimension of the necessity of logical discrimination can the importance of the model of two-state switching devices be seen, and (taking a further step into the empirical data) that the signification of an empirical-physical mechanism as the trigger-mechanism, involved in the firing of the neurons, can be recognized.

Many authors in the field of methodology have insisted on the importance of a certain 'logic' which must precede any fact-finding. This 'logic' can have a loose form such as awareness of certain distinctive features guiding a process of observation, or as a heuristic horizon delimiting a field of investigation, but it has always a certain discriminating power. This capacity for discrimination can be found already in animal nervous systems and in animal behavior. The human body develops high capacities for acquiring and organizing information. Man is capable of constructing varied networks of rules for organization, developing variable symbolic behavior, and responding thus to the innumerable dimensions of the surrounding world that are being disclosed in these ways. These rules are more or less adequate media for human response to and control over reality – and this concerns not only rules for vital and social behavior, but also, as J. Piaget among others demonstrates in his elaborate research, the rules of symbolic logic and mathematics can be considered as patterns of organization.[29]

All this does not imply any reductionism or psychologism. Methodological interpretation precedes empirical description. Husserl's analysis of the human body remains of great importance. It has been explained, in the first part of this article, how Husserl described bodiliness within the methodological scope of intentionality. Here a dynamic structure became visible: that of the interpretive and constitutive activity of subjectivity. There is no metaphysical dualism of a spirit in a bodily box; there is rather the more logical relationship of implication, not physical or causal, between the empirical-physiological components and the intentional structure of bodiliness.

A similar type of analysis in a direction leading away from the dimension of the Ego is, however, necessary. The human bodily existence is not merely a declension of a transcendental subjectivity, but also a response to transcendentally analyzable dimensions of the surrounding world.[30] Processes of logical constitution must be interpreted as processes of interaction; the dynamic structure of the human bodily existence shows interrelatedness to be its fundamental feature; and likewise in its phenomenological respect. An investigation, along the lines of Husserl's program, is here possible. It would make it clear that symbolic information, to which man responds in his cultural and social behavior, is not contained in the objectively given material, like things in a box. There is a relation of implication, as the human evaluations imply certain, methodologically analyzable dimensions of reality. A field of traditionally metaphysical assertions can be brought within the methodological boundaries of various disciplines, if these disciplines remain in contact with a more interdisciplinary, integrating investigation. The phenomenology of the body would be such an integrating type of philosophy.

Husserl's dilemma, resulting ultimately from his concept of a 'lonely'

Ego, could be solved in this way. It is phenomenologically incorrect to reserve the dimension of dynamic subjectivity for the Ego and to omit interrelationship and response to the empirical field or to the natural ('mundane') reality. The Ego (that is the logical dimension of subjectivity) is not lonely, as Husserl states himself that "consciousness is not conceivable without any organization". Organization implies control as well as adaptation – modern biological sciences give us illustrations of this. Intentionality, to be found in scientific knowledge or in bodily orientation or in acts of the will and of consciousness, has two aspects: the activities of construction and of response. Such a phenomenology of the human body would remain a transcendental analysis but would be, at the same time, more involved in the researches of both empirical science and the foundations of logic, as well as in the responsive character of historical man in his social and ethical world.

Husserl's phenomenology of the body, within his program of the transcendental phenomenology, is not an end but a starting point. The task of such a phenomenology is:

(a) to take into account the material presented by phenomenological psychology and by the analysis of ordinary language, as far as these are related to bodiliness;

(b) to give a transcendental, that is, a methodological and structural, clarification and interpretation of this material;

(c) to come into close contact with the methodological results of scientific research into human symbolic behavior and of the biology of the human body – results that often need interdisciplinary investigation;

(d) to give a methodological analysis of organizational structures, relating phenomenological logic investigations (i.e. transcendental philosophical ones) to contemporary symbolic logic, especially to the investigation of its foundations;

(e) to integrate a phenomenological investigation of bodiliness into the wider horizons of human evaluative behavior and of social and ethical responsibility.

University of Leyden

REFERENCES

[1] R. M. Zaner, *The Problem of Embodiment*, The Hague 1964.
[2] C. A. van Peursen, *Body, Soul, Spirit*, Oxford University Press, 1966.

[3] E. Husserl, *Logische Untersuchungen*, II, 1; Halle (1901) rev. ed. 1913, pp. 165, 166.
[4] E. Husserl, *Ideen*, II. The Hague 1952, pp. 152, 153.
[5] *Ideen*, II, pp. 153–157.
[6] *Ideen*, II, pp. 157–172.
[7] E.Husserl, *Phänomenologische Psychologie* (from 1925), The Hague 1962, pp. 196–199.
[8] E. Husserl, *Formale und transzendentale Logik*, Halle 1929, pp. 221–225.
[9] E. Husserl, *Logische Untersuchungen*, II, 2, pp. 196–199.
[10] E. Husserl, *Zur Phänomenologie des inneren Zeitbewusstseins*, The Hague 1966, p. 75 (orig. ed. p. 429).
[11] E. Husserl, *Die Krisis der europäischen Wissenschaften und die transzendentale Phänomenologie*, The Hague 1954, pp. 185–193.
[12] f.i. G. Brand, *Welt, Ich und Zeit*, Den Haag 1955. – E. Fink, 'Die Phänomenologische Philosophie E. Husserls in der gegenwärtigen Kritik' in E. Fink, *Studien zur Phänomenologie*, Den Haag 1966. – K. Held, *Lebendige Gegenwart*, Den Haag 1966. – L. Landgrebe, *Der Weg der Phänomenologie*, Gütersloh 1963.
[13] E. Husserl, *Erste Philosophie*, II, The Hague 1959, pp. 386, 387. – E. Husserl, *Formale und transzendentale Logik*, p. 245.
[14] E. Husserl, *Analysen zur passiven Synthesis* (from 1918–1926), The Hague 1966, pp. 216, 377–381.
[15] L. Wittgenstein, *Tractatus Logico-philosophicus* (the first translation, London 1922, is being quoted) 5.631–5.633a; 5.641.
[16] L. Wittgenstein, *Zettel*, Oxford 1967, Nr. 479–483. L. Wittgenstein, *Philosophische Bemerkungen*, Oxford 1964, Nr. 70–74.
[17] L. Wittgenstein, *Zettel*, Nr. 614.
[18] L. Wittgenstein, *The Blue and Brown Books*, Oxford 1958, pp. 46–48, 63–67.
[19] L. Wittgenstein, *Philosophical Investigations*. Oxford 1953, Nr. 404–410. – L. Wittgenstein, *The Blue and Brown Books*, pp. 68, 69.
[20] S. Hampshire, *Thought and Action*, London 1959, pp. 85–90, 145–168.
[21] The following titles can be mentioned: L. Brillouin, *Science and Information Theory*, New York 1956. – W. M. Elsasser, *The Physical Foundation of Biology*, London 1958. – H. Quastler, *The Emergence of Biological Organization*, New Haven–London 1964. – Chr. P. Raven, *Oogenesis; The Storage of Developmental Information*, Oxford 1961. – E. Schrödinger, *What is Life*? New York 1945.
[22] The following works on this theme give an orientation: J. P. Schadé and D. H. Ford, *Basic Neurology*, Amsterdam 1967. – N. Wiener and J. P. Schadé (eds.) *Cybernetics of the Nervous System*, Amsterdam 1965. – D. E. Wooldridge, *The Machinery of the Brain*, New York 1963.
[23] W. H. Thorpe, *Biology, Psychology and Belief*, Cambridge 1961. – W. H. Thorpe, *Biology and the Nature of Man*, London 1963.
[24] M. Polanyi, *Personal Knowledge*, London 1962 (2nd ed.), p. 377.
[25] R. Ruyer, *L'animal, l'homme. La fonction symbolique.*, Paris 1964, pp. 44–55.
[26] J. Ladrière, 'Limites de la formalisation', in J. Piaget, *Logique et connaissance scientifique*, Paris 1967, pp. 312–333. – A general introduction: E. Nagel and J. R. Newman, *Gödel's Proof*, London 1959.
[27] P. W. Bridgman, *The Way Things Are*, Cambridge, Mass. 1959, pp. 1–11. cf. f.i. G. Frey, *Sprache – Ausdruck des Bewusstseins*, Stuttgart 1965.
[28] E. Husserl, *Logische Untersuchungen*, I, pp. 67–69.
[29] J. Piaget, *Biologie et connaissance*, Paris 1967, pp. 350–381.
[30] C. A. van Peursen, *Facts, Values, Events* (English translation in preparation).

RICHARD SCHMITT

ON KNOWING ONE'S OWN BODY

For most philosophers the only problem about the body concerns its relation to the mind. Not so for a transcendental philosopher like the later Husserl. He claims some sort of relativity of the world to a subject. The world is said to be 'constituted' by the subject. Many, if not all, of these subjects are embodied. The transcendental philosopher must therefore ask himself whether the body functions in the process of constituting the world. But the body is also in the world and is, in many respects, like the other entities in the world. Thus the body is also constituted by what is relative to a subject. The transcendental philosopher must make clear in what sense the body is constituted and participates in the process of constituting. More specifically, he must make clear in what sense it is true that any particular subject, say I myself, constitutes any particular body, for instance my own, and does so in a different way from that in which it constitutes any other person's body.

The notion of constitution is intimately linked to the notion of experience. Individual things, as well as the world as a whole, are often said to be constituted in experience. The same is true of the body. An account of the constitution of the body thus involves an account of one's acquaintance with one's body. The transcendental philosopher who wants to clear up questions about the constitution of the body must make clear in what ways our own bodies are familiar to us.

In this paper I want to raise four difficulties about Husserl's account of constitution, particularly with respect to the body. I want to show that the source of these difficulties lies in the inadequacies of Husserl's conception of the experience of one's own body. Some hints for an alternative account of this experience are provided by Husserl. They point in the direction in which we must look for a more adequate explication of bodily self-knowledge and, thus, of the constitution of the body.

Tymieniecka (ed.), Analecta Husserliana, Vol. I, 152–169. *All Rights Reserved*
Copyright © 1970 *by D. Reidel Publishing Company, Dordrecht-Holland*

I

It is the central thesis of transcendental phenomenology that the world is constituted by a community of transcendental egos. The precise sense of 'constituted' is extremely difficult to untangle, but the claim that the world is constituted implies at least that the existence of the world and its existence, in just the way it does exist, is dependent on human subjects where 'dependent' does not mean 'causally dependent'. A transcendental philosopher does not merely assert, what no one would deny, that human beings affect the world in accord with causal laws. Husserl puts the transcendental claim as follows:

> The objective world ... with all its objects derives ... its entire meaning and its claim to existence from me, as the trancendental ego ... (CM 65).[1]

The transcendental ego is a consciousness. Consciousness is intentional. The claim that the world is constituted thus means that the existence of the world, and what it is like, is dependent on the intentional acts of conscious beings like myself. It appears to the transcendental philosopher that the occurrence of intentional acts is at least logically necessary for the existence of the world. If we ask, on the other hand, for the characteristics of these intentional acts, we receive a very different account. It now seems that intentional acts are anything but necessary for the existence of intentional objects, either singly or as the totality, which we call 'the world'. On the one hand, Husserl tells us that percepts are distinct from intentional objects:

> The *tree as such*, the thing in nature, is anything but the *percept of the tree as such* which belongs inseparably to any given perception as its meaning. The tree as such may burn ... the meaning of this perception ... does not burn. (Id I 222)

Not only are percepts and what is perceived by means of it utterly different, but there may be a percept where there is no corresponding object perceived (Id I 223). The converse is obviously true: there may be a tree and no corresponding percept and act of perception (Id I 92). Intentional acts are described in such a way that the intentional object is not dependent on intentional acts either for its existence or for being just the sort of thing it is.

What is true of individual acts and their relation to their intentional objects is true of the totality of my intentional acts in relation to the world as a whole. The world may be destroyed, while my acts continue to exist (Id I 114ff). It may, on the other hand exist without being perceived (Id I 92–133). The world as a whole is logically independent of the intentional acts of a particular consciousness.

It now seems as if transcendental phenomenology, at least in the version of it that Husserl develops, is caught in a contradiction in that it both affirms and denies that the existence of the world is logically dependent on the occurrence of acts of consciousness. In giving a minimal elucidation of the concept of constitution, Husserl seems to tell us that the occurrence of intentional acts is a necessary condition for the existence of the world. In elucidating the intentional character of acts of consciousness he denies that logical dependence.

This problem, as stated, applies to any intentional act and any intentional object. It therefore applies to the more specific range of intentional acts involved in knowing one's own body and the corresponding intentional objects which are under discussion in this paper.

II

Material objects are constituted in experience. When Husserl comes to talk of their constitution, he points out that the distinction between an apparent thing and a real thing, between what appears to us as material thing and what really is a material thing, depends on the relations of any one thing to the context in which it is experienced. If we pay attention only to the perceptions of a given thing, we are unable to discriminate between real and apparent things, because we are not able to discriminate changes in the thing from changes in our perception of them. Thus to know a real thing is to know its causal connections with other things (Id II 41–45). We do not know the real properties of a thing, unless we are able to distinguish those changes in our perception of it, which are perceptions of changes in the thing itself, from those which are changes in our perception of the thing.

Prominent among the causes of changes in the perception of a thing are obviously changes in the position and condition of the perceiver. Thus among the items of information about the causal context of a given thing are bits of information about the perceiver and relevant changes in him. Accordingly the constitution of a material object in perception involves kinaesthetic sensations (Id II 55ff). Whether a given change of size or shape is a change in the object or a change in my perception of it depends on the constancy or change of my position in relation to the object perceived. I am informed of these circumstances by my kinaesthetic sensations.

All of this seems plausible, but the account given is incomplete; more needs to be said about the kinaesthetic sensations. The difficulty which leads Husserl to claim that "real properties are *eo ipso* causal ones" (Id II 45) applies also to human bodies. The kinaesthetic sensations alone cannot inform me about the position of the body since they might, and sometimes are, misleading. Think of the phantom feelings of the amputee. Before I can count on the kinaesthetic sensations as reliable information about my movements, I must have constituted my own body as real, situated in space, and thus as causally connected with other things, e.g. the floor on which I stand and on which my feet push when I walk.

The question about the constitution of material objects leads us back directly to questions about the constitution of the human body not only because the body is an instance of a material object but also because the constitution of any material object involves reference to my kinaesthetic perception of my body and thus the real existence of that body.

Bodies are constituted in sensation, specifically in tactile sensations: "A subject gifted only with sight ... could not have a body" (Id III 122). By tactile sensations Husserl means sensations of touch but also sensations of temperature, taste and smell. These sensations fulfill their constitutive rôle since they are, at the same time, sensations of objects touched and sensed and sensations of the body that does the touching or sensing. In touching a thing, I feel the surface of the thing touched and the fingers that do the touching (Id III 123).

But this account is full of difficulties. The constitution of material objects requires, as we saw, reference to the perceiving body, specifically to kinaesthetic sensations which enable us to differentiate those sensations that inform us about the properties of the objects perceived from those that are only results of bodily movements and thus tend to mislead us about objects. My body is also a material object, and the difficulties that arise in the perception and constitution of other material objects apply also to the constitution in experience of my body. Some sensations inform me about my body; others mislead. Errors arise in the case of phantom feelings or in drugged states where my body appears bloated beyond all normal size or shrunk to a fraction of normal size. Here the boundaries of my body do not seem to be what they are ordinarily. Similarly, in a heightened state of sensitivity, I may feel pain when a door is slammed, and it may appear to me that my nerves extend into the door frame. If the body is fully constituted, we can readily differentiate these sensations from normal ones as being drug induced. But we are told by Husserl that these tactile sensations are constitutive of the body. We are not told, however, what there is in the process of constitution that allows us to differentiate between sensations that are normal and those that mislead. There is nothing in the constitution of the body that fulfills the function of kinaesthetic sensations in the constitution of other material objects namely, to eliminate misleading sensations from the range of constitutive ones. Thus it is not at all clear how the body gets to be constituted, since it would seem that it already needs to be constituted, if we are to discriminate reliable from unreliable sensations. The process of constitution of the body, it would seem, cannot get started.

This is a genuine problem and one of which Husserl is aware. The body is constituted in tactile sensations of things, but these tactile sensations of things already presuppose that the body itself is a thing, i.e. is already fully constituted (Id III 121). Husserl makes no suggestion for a solution of the difficulty except to say that nature, the world of things other than our body, and the body are constituted together and mutually related to each other (Id II 124). That remark, however, is not a solution of the problem but a restatement of it. The fact remains that constitution of things and of bodies is intimately linked to the experience of things and bodies. But not all experiences are reliable. Constitution in experience calls for a way of discriminating reliable from misleading sensations. There is such a way if the body is fully constituted. But no such discrimination can be made, it seems, in the process of constituting the body itself.

III

An additional difficulty of Husserl's account of the constitution of the body can be stated most briefly by saying that Husserl may have told us how the body or a body are constituted, but he has not told me how I constitute *my* body. His claim is that the body is constituted from tactile sensations which include kinaesthetic sensations, sensations of warmth, wetness etc. The difficulty as to how I constitute *my* body is partly concealed by liberal use of the notion of kinaesthetic sensations, since these presumably are sensations which I have of my own body, and thus, we may these, my body is constituted by me as I come to know it by means of these kinaesthetic sensations. But this account would be adequate only if what I feel in a kinaesthetic sensation, its 'content', were to contain a reference to *my* body. But it does not, of course, contain such a reference. When moving my arm I have certain muscular sensations, but it is not contained in these muscular sensations that it is *my* arm that is being moved. When we call these muscular sensations, and others like them, 'kinaesthetic' we already know that these are sensations of goings on in my body, but this knowledge is not derived from the kinaesthetic sensations themselves, at least not from the kinaesthetic sensations alone.

It is true, of course, that I have kinaesthetic sensations only of *my* body. But in order to know that I must already have identified my body, and its boundaries, as mine. Without such an identification I do not know that the muscular sensations in my body are all sensations in one and the same body.[2]

IV

Husserl speaks of the human body most frequently in cognitive contexts where he discusses the rôle of that body in the constitution of objects of perception. But we use bodies not only for perception. Bodies are also the vehicles of actions. If there are difficulties in understanding how the body can be constituted in experience and as "organ of perception" (Id II 56), there are distinct but no less pressing difficulties in understanding the constitution of the body as instrument of action.

Husserl links, as we have seen, the notion of constitution with that of experience. The body, like the rest of the world, is constituted in experience. My body is constituted by me in my experience. But having an experience always involves the ability to say that one is having an experience and to give a description of its content. This is clearly implicit in Husserl's analysis of intentional acts. Every act has, among other elements, a content, called a 'noema', and Husserl insists that the noema of any given act can, in turn, become the intentional object of another act. It is always possible to say what it is one is experiencing (Id I 222). It would appear therefore that the domain of the constituted is coextensive with the domain of what is experienced, and that what is experienced is something of which we are, in principle, able to give a description. But the body, insofar as it is the instrument of action, is not in this way accessible to us. If being experienced, in the sense discussed, is a necessary condition for being constituted, then the body as organ of perception may be constituted, but the body as instrument of action is not, for actions are such that there is a good deal about them that I am not able to describe. Take the cases where I perform more or less habitual actions without being fully aware of them. Examples are the nervous habits, twitches, facial expressions with which one accompanies lectures or conversations. One is often so little aware of these habits that one is surprised if told that one has them. Here I do not know, say, that I have a habit of pulling my earlobe when thinking about something. I therefore am in no position to describe that action. On Husserl's view that action is not an element in my experience.[3]

Husserl, quite characteristically, speaks of my kinaesthetic sensations under the heading of the constitution of the body as "organ of the will" (Id III 119–120). But the body as executor of my choices and decisions is only very fragmentarily experienced in kinaesthetic sensations. Take now a case where I am explicitly aware of what I am doing. I am reaching out to open the door. But in this case also there is a good deal going on in my body while I am acting of which I am not aware. I am not even thinking here of the physiological process, firing of nerve fibers and contraction of specific configurations of muscles, of which opening a door consists. When opening a door I am not readily able to describe the precise trajectory of my hand, the path it describes in space. I know even less about the way in which the balance of my body shifts as I lean forward, or how I shift my weight from one leg to the other. The gross bodily motions involved in performing an action like opening a door are not experienced by me. To the extent that constitution takes place in experience, and the body, as agent, is not experienced, the body as agent is not constituted. But this conclusion is clearly incompatible with the central thesis of Husserl's transcendental phenomenology that the world, and whatever it contains, including our own bodies, is constituted by transcendental consciousness.

V

The point of raising these four difficulties was not just to find fault with Husserl's discussion of constitution and of constitution of the body. The point was rather to get some suggestions about ways in which we must *not* talk about constitution and to get some, rather more vague, suggestions about how we might talk about constitution more adequately. Having learned something about the conceptual machinery needed in order to make clear what the transcendental philosopher is telling us when he talks about constitution, we can, in addition, turn again to Husserl's texts and find that he was to some extent aware of the ways in which constitution must be spoken of, if we are to make clear what we mean by it. The preceding criticisms thus serve to make general philosophical points and to throw further light on Husserl's views.

The first three difficulties served to make the necessary negative points. We learn from them how not to talk about constitution, for the difficulties that we encountered can be shown to result from a particular explication of the concept of consciousness. If the concept of consciousness were correctly explicated in this way, the claim that consciousness is constitutive could not be rendered intelligible. The fourth difficulty yields some suggestion for an alternative explication of consciousness.

Consciousness, as everyone knows, is said by Husserl to be intentional. The concept of intentionality is one to which Husserl devotes a good deal of close analysis. One of the results of these analyses is that all intentional acts are propositional. Every intentional act has, according to Husserl, a content, called a 'noema'. The noema, or at least important elements of it, is a '*Satz*', a proposition. Husserl insists that he is not using the word '*Satz*' to refer to a linguistic entity. I take him to deny thereby that the noema is either a sentence, or a statement. It is the sort of logical entity which philosophers traditionally have called propositions (Id I 324).

Consciousness, in its different modes of believing, wishing, valuing, expecting, ordering etc., is always to be understood as different relations to a proposition. I can regret to-day what I was looking forward to yesterday. The two acts differ with respect to the relation that I sustain to one and the same proposition on the two different occasions.

Propositions are true or false and that objectively so. The truth of a proposition is not affected by its occurrence in any particular intentional act, by being related in a particular way to a particular person, by being expressed in a particular language. The same is true of the states of affairs by reference to which a given proposition is true. That state of affairs obtains, as it obtains, regardless of any intentional acts of which it is the intentional object.

If we adopt this explication of intentionality, and thus of consciousness, we are not able, as we have seen, to assert that consciousness is constitutive without generating logical difficulties. But Husserl's dominant explication of intentionality does not only have the weight of tradition on its side, it also applies extremely well to a certain range of cases. Knowing facts about the world, both in everyday affairs and in the sciences, natural and otherwise, seems to be propositional just in the way that Husserl describes. What is questionable, what in fact we are now seeing to be a serious error is the implicit claim of Husserl's (and of many other contemporary philosophers) that all acts of consciousness must be explicated in the same way as theoretical knowledge is to be explicated. Husserl assumes, it seems, that knowing facts and laws about the world is the model for all acts of consciousness. For a transcendental philosopher that is not a tenable position.

This point is supported by the second difficulty and the question concerning the constitution of the body. The body is both perceived and organ of perception. It is the latter – organ of perception – in a very different sense from that in which, say, my heart or my liver are said to be organs. For my heart and my liver function whether I am aware of them and their functioning or not. The same is true of my retina or my iris. It is not true of my eyes. Husserl's discussions of the function of the body in perception of the world are extremely interesting for their insistence that awareness of the body as organ of perception is a necessary ingredient in the experience of anything.

It is clear, however, in the light of the second difficulty, that we cannot construe all bodily self-perception on the same model on which we construe perception of entities other than our bodies. For if perception of things other than my body requires perception of the body and the same applied also to perception of that body itself, we would be led into an infinite regress where any perception of the body would presuppose other perceptions of the body. It follows that the way in which I perceive my body must be different, at least in some cases, from the way in which I perceive other things.

Traditionally, it had been thought that the difference between perception of my body and my perception of all other things could be drawn in epistemological terms, because my self-perceptions are infallible whereas my perceptions of the external world are not. If this were true, any given perception would not need to be supported by other perceptions of the body, since such support is needed only where evidence for the reliability of any given perception is required. But no evidence is required for infallible perceptions. Hence the regress is avoided if we regard my perceptions of my body as infallible.

But this solution is, of course, unacceptable, because my perceptions of my body certainly are not infallible as shown by the existence of phantom feelings. The infinite regress with which we are threatened when we construe perceptions of one's own body as the same kind of perceptions as one's perceptions of all other things, must be avoided by arguing that self-perception of the body differs in other ways from perceptions of external objects. It is not too difficult to see that the difference consists in this that self-perceptions, in many cases, are not what Husserl calls intentional acts; they are not to be construed as specific relations to propositions.

Individual perceptual acts are not adequately evident even if they are apodictic (CM 62). Given any particular perception, we may, for instance, state what it is we perceive, but the statement we make here is no more than a perceptual claim. Further perceptions, my own or someone else's, may invalidate that first claim. Perceptions are defeasible. This presumably also applies to perceptions of the body for the reasons cited. Kinaesthetic sensations certainly are not incorrigible.

Propositions are either true or false. Intentional acts, if construed as propositional, thus are relations to what is either true or false. If we, besides, agree that perceptual acts are corrigible, we are led to conceive of perception as a series of acts because any given perceptual act needs the evidential support of other acts before it yields justified belief. If we construe our bodily self-knowledge as propositional in his way, an infinite regress is generated, because the body perceived is also the organ of perception. If we accept the propositional analysis of perception, we are committed to the view that any given perception of the body needs to be supported by other perceptions of the body. But these perceptions of the body are feelings *in* the body. They therefore cannot have evidential value unless we have good reason to think that we have a body, to begin with. But these same perceptions, which have evidential value only if I already know that I have a body, are intended to provide evidence for just that belief. Thus it would seem that I come to know my body *via* my perceptions of it only if I already know that it exists, but if it is also held that I know that my body exists because I perceive it, then no perception of my body, can ever give evidential support to any other perception and thus we cannot begin to perceive our own body.

VI

The third difficulty concerns the constitution of my body as *mine*. Granting for the moment that I constitute my body in the course of becoming familiar to it by means of kinaesthetic sensations, it is not by any means clear how I know from these kinaesthetic sensations that it is my body that I know and, thereby, constitute.

This difficulty also arises from construing all acts of consciousness as intentional in the sense of propositional acts. There is, of course, good reason to think that a good deal of my awareness of my own body can be construed propositionally. If I touch a particular spot on my body, which had been bruised, to find out whether the spot is still tender, the act of feeling has all the ingredients of an intentional, propositional act. There is the act itself, the noesis, namely feeling. There is the sensation I have, a feeling of soreness, there is what I take that feeling to be, the noema, the peculiar soreness of a place on my body that had been bruised. Finally there is the intentional object, the bruise, which may or may not exist. But when I feel the bruised spot, I do not only have a sensation of that spot, I also am aware that I am feeling the spot with my fingers. I am aware of my body as felt. I am simultaneously aware of my body as doing the feeling, as I am not, for instance, if the doctor touches the bruised spot to find out whether it still hurts. Is this awareness that I am feeling my own body (or another body for that matter) with my own fingers to be thought of as itself a propositional act? The third, like the second difficulty, suggests that we must not construe this awareness of what I am doing with my body in this way.

Intentional acts, on Husserl's standard analysis have several elements. There is the I, the noesis, the sensation (at least in certain classes of acts, namely perceptual ones) the noema and, sometimes, the intentional object. Husserl arrives at this list of elements of intentional acts by his usual method of free imaginative variation according to which we can see that any given 'I' may perform a large range of noeses on a large range of sensations with a similarly large variety of noemata. In all of these cases we may speak of intentional acts. All that is needed is that I perform some noesis with some sensation and some noema. There is no necessary connection between any particular 'I' and any particular set of sensations, noeses, or noemata.

If my awareness of myself as performing certain acts of feeling are to be construed on this model, there is no necessary connection between any set of acts involving kinaesthetic sensations of my acts of feeling and the 'I' which occurs in all of these acts. This would suggest that it is a contingent fact that these kinaesthetic sensations are sensations of my body. It would not be impossible that I should be having the identical kinaesthetic sensations of someone else's body. But that surely is impossible. It is not only false as a matter of fact that I have kinaesthetic sensations of your body, it is false by virtue of the meanings of terms. But if that is true, there must be a necessary connection of some sort between the kinaesthetic sensations and my 'I' which could not exist if the awareness of my own body *via* kinaesthetic sensations were to be construed on the propositional model of intentional acts. It follows that when I feel the bruised spot on my body, my feeling the body differs from my awareness that it is I that is doing the feeling, it does not only in that the two acts have different sensations, noemata and intentional objects, but in that the two acts have an utterly different structure, in that my awareness of my own body is much more intimately related to the 'I' than any other sensation that I may have.

This is not to deny, of course, that I do perceive my body. Nor is it to deny that there are kinaesthetic sensations and that perceptions of my body *via* kinaesthetic sensations are, like all other perceptions, corrigible. What we must say, on the contrary, is that my awareness of my body and what it does is not to be thought of as exclusively mediated by kinaesthetic sensations. The discussion of the fourth difficulty will suggest the outline of an alternative way of construing this awareness of my own body.

If we approach the text with this issue in mind we find that Husserl is not very clear on this point. It looks as if some of his remarks supported the view that feeling that I am touching, for instance, is itself to be construed on the propositional model. Other remarks seem to support the opposite view. On the one hand, we are told that sensations of touching (felt for instance in my fingertips when I am touching something) – Husserl calls them '*Empfindnisse*' – are apperceived (*aufgefaßt*) in intentional acts in the same way in which sensations enter into acts of perception, for instance, seeing the apple tree in bloom in the garden. Husserl provides a propositional analysis of the latter. The present passage suggests that he envisages the same analysis for sensations in my fingertips when touching something. (Id II 151–152). But there are other passages that seem to point to a very different scheme of analysis for *Empfindnisse*. We are told, for instance, that the distinction between the sensations I have when touching something, on which I rely when I say that the touched surface is, say, rough and cold, cannot be separated from the sensation I have in my fingertips when touching. The two sensations can be separated in thought only (Id II 155–156). Nor are they sensations of the same sort. The awareness I have of my fingertips doing the touching and feeling a rough and cold surface is not itself an awareness of a thing. It does not admit of the difference between what seems and what is, it is not, in Husserl's terminology, '*abgeschattet*'. It does not belong in a '*sinnliches Schema*'. That means that it is not a possible perception of a thing (Id II 150).

When Husserl tells us that the sensation of the rough and cold surface can be separated in thought only from the sensation of moving my arm to do the touching, he is implicitly denying that the latter awareness is a perception of the same sort as the former. For he is telling us that there cannot be an act of, say, touching my skin unless there also is a *veridical* awareness of this act. But perceptual acts, we saw earlier, are not related in such a way that, given the occurrence of one act, another act which cannot be defeasible must also occur. Similarly, perceptions are perceptions of something. To say that perception is '*abgeschattet*' and to talk about a '*sinnliches Schema*' is to say that any given thing is perceived in a series of acts and that the beliefs which any single act yields must be supported by each other before we have good grounds for believing that we perceive a certain thing. But my awareness of moving arm and fingers when I touch my skin is not of this sort. The awareness does not have to be supported by other acts of awareness. In it there is not a thing or an event that is being perceived. I am simply aware of *what I am doing*. Being aware of what I am doing is of a very different nature from perceiving what someone else is doing.

VII

The preceding section has made a number of negative points about bodily self-knowledge and about the constitution of the body. We must now look at the fourth difficulty to find some, admittedly, meager hints as to the proper way for construing bodily self-knowledge and the constitution of the body.

The fourth difficulty consisted of this: If we link the notions of constitution and experience and then explicate 'experience' as a variety of different relations to propositions – perceiving, seeing as, inferring, being evidence for etc. all being different relations to propositions – and if we further agree that being related to a proposition goes with being able to state that proposition, then the body as instrument of action is not constituted, since a good deal of what we do is such that we cannot talk about it at all, or only very fragmentarily.

In the light of the difficulties raised earlier, this suggests that we should explicate experience not as propositional but as more like acting, than traditional philosophers have been willing to do. It is immediately obvious that that is a startling and, at first, repugnant suggestion. For it has been thought for a long time that experience was either the antecedent of, or a case of knowing (or at least of believing), while acting was perhaps a causal, but not a logical antecedent of knowing and believing and certainly not an instance of it. Acting is either doing or making, experience, it has been thought, is neither acting nor making. This sharp separation between knowing (or believing) and acting is a direct consequence of the traditional analysis of believing as propositional. If believing is being related to a proposition, then surely acting is not a case of believing although, of course, one may be acting *on* a belief. Acting, in the most obvious cases, involves bodily motions, but it does not involve anything that can be said to be true or false. Propositions, on the other hand, are true or false, but they do not involve bodily motions. As long as we insist on the propositional analysis of conscious acts, we may say that actions are caused by conscious acts but not that they are conscious acts. If, on the other hand, we do want to say that some actions *are* conscious (not only that they are caused by conscious acts), we must surrender the uniform propositional analysis of conscious acts.

This suggestion gains plausibility if we apply it to the particular subject under discussion, namely bodily self-knowledge. We saw in the preceding sections that Husserl regards feeling my body as central to the knowledge I have of my body and thus to its constitution. We also saw that it is not plausible to give a propositional analysis of this feeling of my own body. What we saw instead was that feeling that I am touching a certain surface is an awareness of what I am doing. Thus the proper analysis of bodily self-knowledge will perhaps emerge if we can provide an account of touching, feeling, looking for, investigating where these are understood either as full-fledged activities or as episodes which are in important respects like activities.

If such an analysis of touching, feeling, looking for is to help in resolving the difficulties stated earlier regarding Husserl's account of constitution, in general, and of the constitution of the body, we must begin by showing that acting in the senses relevant in the present context is to be understood as a conscious activity. That means that acting must be understood in such a way that a person who acts, in some sense, must know what he is doing; otherwise he cannot be said to be acting. But furthermore this sense of 'knowing what one is doing' must be such that it is not capable of a propositional analysis. Or rather we have already seen that knowing what one is doing is not propositional. We need here a sense of knowing that is not propositional.[5]

The fourth difficulty does not yield any firm suggestions as to a resolution of the problems stated, but it does suggest some desiderata for any analysis of bodily self-knowledge which is adequate for an explication of the thesis that bodies are constituted by the subject that inhabits them:

(1) Some mental acts, like feeling, touching, looking for are to be thought of as actions.

(2) Actions are to be analysed as conscious goings on, in the sense that knowing what I am doing is a necessary condition for some episode being an action.

(3) Actions, although conscious, are not to be thought of as involving a relation to a proposition.

The third desideratum rules out the account of action according to which acting is being related to the special sort of proposition that we call 'a rule' and this mental episode, in turn, is causally related to certain bodily motions.

An analysis of, say, touching which satisfied these desiderata – if such an analysis can be given – would serve to resolve the fourth difficulty in a fairly straightforward way. For acts like touching would now be given an analysis where they would count both as instances of an action, and of awareness, for the sense of acting that is needed here is such that acting involves knowing what one is doing. In that case, we could continue to link the notion of constitution with that of experience and yet hold that the body as agent is constituted.[6]

The third difficulty, as to how an explanation can be given of the constitution of *my* body, comes closer to resolution if we adopt this new pattern of analysis, at least for some acts. As long as we insist that all conscious acts are propositional, I am merely contingently related to all my acts and thus to whatever is constituted in them. But if at least some actions are to be counted as conscious acts, albeit not propositional ones, in that the action involves on my part an awareness of what I am doing, then there is a certain awareness which I have only of my body, since it is only my body which is an agent in my actions. In this way I am aware of myself in ways in which I am not aware of anyone else, and in this awareness the body is constituted as mine.

If this analysis can be worked out, it is immediately clear, that the second difficulty, which requires that my awareness of my body be different from that of other material objects, has found a solution. The infinite regress which threatens Husserl's dominant account of the constitution of the body can also be avoided, for on this new analysis of some conscious acts my awareness of my body is at least in part non-propositional and thus does not require evidence. Thus knowing that I have a body is not something which I need to provide evidence for. My awareness of my body can therefore provide evidence for propositions about my body without threat of circularity.

In order to avoid the first difficulty, that a propositional analysis of consciousness is logically incompatible with the thesis that consciousness is constitutive – we need to lay down and satisfy a fourth desideratum for any adequate analysis of bodily self-knowledge:

(4) The non-propositional sense of experience, including experience of one's own body, is logically prior to the propositional sense of experience.

If we can show that to be true, we can resolve the first difficulty. Unless we can satisfy these desiderata with our account of bodily self-knowledge or an alternative set of desiderata that also avoids the four difficulties, the transcendental thesis as applied to one's own body is untenable.

VIII

Our reflections about the problems of constitution and of constitution of the body enabled us earlier to see two divergent strands in Husserl's reflections about the genesis of the body. We are now in a position to apply what we have learned to some other texts to show, once more, that Husserl's analyses of bodily self-knowledge and constitution move in sharply differing and, in fact, irreconcilable directions. This is obviously not intended as a criticism of Husserl, since he was the first to insist that his analyses are provisional.

We have seen that Husserl associates constitution with intentionality and explicates intentional acts as relations to propositions. Having seen that there are possibly alternative analyses of the conscious acts which may be said to constitute, at least, one's own body, we can now also find passages in Husserl where he seems to recognize the possibility of such alternative analyses. There are passages like this one:

> The body is not just a thing; it is an expression of the spirit *and it is, at the same time, an organ of the spirit.* (Id II 96.)

On the propositional analysis of consciousness, being conscious of something is to be related to a proposition which may be true but may also be false of the intentional object. Being related to the proposition does not assure one that the intentional object exists. If we conceive constitution on this model, the body which is constituted by a subject is an intentional object for that subject. On that analysis of constitution of the body, it is possible for a subject to exist without a body. But Husserl quite explicitly denies that in the passage under discussion. There can be, he thinks, a body that harbors no mind. "The body, on the other hand, cannot disappear" (Id II 94). The sense in which the body is said to be an organ of the spirit is such that having a body is essential for a spirit. But if that is true, we must either deny that the body is constituted in experience or reject the propositional analysis of experience. There are excellent reasons for thinking that, if clearly confronted with these alternatives, Husserl would have chosen the latter. We find, thus once again, that the propositional analysis of experience and of bodily self-knowledge is not uniformly acceptable to Husserl himself. It was however sufficiently dominant in his thought that the alternative analysis was never worked out.

Brown University

REFERENCES

[1] I am citing these works of Husserl's in this paper: *Cartesianische Meditationen, Husserliana,* Vol. I (CM); *Ideen zu einer reinen Phänomenologie und phänomenologischen Philosophie,* Vols. I–III, *Husserliana* Vols. III–V; (Id I, Id II, Id III).

[2] In some passages Husserl places great emphasis on the fact that tactile sensations of my body are double. When I feel my left hand with my right hand, my right hand feels the rough and cool surface of my left hand, but at the same time my left hand feels the smooth and warm surface of the fingertips of my right hand (Id II 144–145.) This observation does not suffice to explain how I constitute my body as mine. In fact, it only serves to make clear how I discover that my two hands are both hands of one and the same body, if I already know of each hand that it is mine. The case presupposes the awareness that it is I who is using my hand to feel something, in this case my other hand.

[3] Habitual actions of this sort, we might say, are virtually unconscious and thus are not to be accounted actions at all. This is implausible because we may be made aware of such actions without changing the actions themselves in any way. It is also implausible insofar as habitual actions of this sort are taken to be expressive. They have symbolic value. As such they must be regarded as actions.

[4] If we construe all consciousness as propositional we may of course use 'constitution' in the sense in which Kant uses the term, to suggest that the world as we know it is not identical with the world as it is in itself, since all our knowledge rests on certain premises not derived from and not confirmed by experience. But Husserl explicitly rejects Kant's distinction between the world as it is for us and the world as it is in itself. He thus also rejects the meaning that Kant attached to the term 'constitution' (Id I 98).

[5] I have made some relevant suggestions in my paper 'Two Senses of Knowing', *Review of Metaphysics* **18** (1965) 657–677.

[6] It would, however, only resolve the question how the body as agent can be said to be constituted. It would not help to clarify the very different claim that Husserl also makes, that the body as material object is constituted by myself. This distinction has been insisted on by Merleau-Ponty. See *La phénoménologie de la perception,* Paris 1945, p. 123 n. 1.

ELISABETH STRÖKER

DAS PROBLEM DER ἐποχή IN DER PHILOSOPHIE EDMUND HUSSERLS

I

Die entscheidende Wende, die Husserl seinem Denken wenige Jahre nach dem Erscheinen der *Logischen Untersuchungen* gegeben hat, ist bezeichnet durch den Begriff der Epoché. Husserl selbst verstand diese Wende als Abkehr von der psychologisch-deskriptiven Phänomenologie im Sinne einer Analyse von Erlebnissen des empirischen Bewußtseins; und er forderte nunmehr eine transzendentale Phänomenologie, welche die Bedingungen der Möglichkeit empirischer Erkenntnis des Seienden aufzuklären – ja welche letzthin den Sinn von Sein und Seiendem schlechthin freizulegen habe.

Unbeschadet des nach wie vor 'phänomenologischen' Anliegens verlangt die transzendentale Untersuchung eine neue Methode, die durch eben diese Epoché eingeleitet wird. Sie soll eine Reihe von Reduktionen ermöglichen, die schließlich in ein neues Feld extramundaner Erfahrung, nämlich des transzendentalen Bewußtseins führen, aus dessen konstituierenden Leistungen das Sein der Welt und mit ihm auch das Sein des empirischen Subjekts in der Welt begreiflich werden soll.

Es kann in dieser kurzen Abhandlung nicht darum gehen, die Husserlsche Reduktionsproblematik in all ihren Verzweigungen aufzunehmen. Sie möchte lediglich einer Forderung nachgehen, die Husserl selbst für seine Philosophie des öfteren erhoben hat: Die eigentümliche Rückbezogenheit der Phänomenologie auf sich selbst bringt es mit sich, daß sie sich auch die Methode, die sie übt, zu voller Klarheit bringen muß; daß sie ferner, gemäß ihrem Leistungsanspruch für die gesamte Erkenntnis, diese Selbstklärung wiederum aus phänomenologischen Quellen zu leisten hat.[1]

Husserl hat diese Aufgabe offenbar in einer Vorlesung des Wintersemesters 1923/24 selber in Angriff genommen. Dort heißt es: "Hier beginnt die Behandlung des Problems, wie Sinn und Leistung der Epoché zu klären sei, also eine phänomenologische Theorie der Epoché auf dem

Tymieniecka (ed.), Analecta Husserliana, Vol. I, 170–185. *All Rights Reserved Copyright* © 1970 *by D. Reidel Publishing Company, Dordrecht-Holland*

Grunde einer Theorie von Reflexion überhaupt".[2] Betrachtet man jedoch dieses Lehrstück genauer, so kann es, gemessen an seiner Zielsetzung, befremden. Husserl expliziert dort an Hand einer Dingwahrnehmung den Vollzug einer Einzelreduktion, also die methodische Zurückführung eines wahrgenommenen Gegenstandes auf das pure Gegenstands*phänomen* im wahrnehmenden Bewußtsein. Er erläutert dabei zugleich die verschiedenen Reflexionsstufen des Ich und gibt anschließend einen Exkurs über den Begriff des Interesses, der Einstellung, des Themas, wobei es zu neuen aktphänomenologischen Differenzierungen kommt in Gesamtakte und Teilakte, Teilakte in dienender und herrschender Funktion, Akte in Haupt- und Nebenintentionen, schließlich zur Unterscheidung aktueller und habitueller Themen. Was Husserl damit als "Teil einer phänomenologischen Theorie der Epoché" gegeben hat, ist also im Grunde eine phänomenologische Analyse derjenigen Akte, an denen die Epoché vorzunehmen ist, nicht aber eine phänomenologische Untersuchung der Epoché selbst.

Damit aber blieb seine eben skizzierte methodologische Forderung bislang ein unerfülltes Programm. Die phänomenologische Methode "aus phänomenologischen Quellen" zur Klarheit bringen, wie Husserl es verlangt, das wird nur heißen können: reflektierend eine zuvor schon geübte Methode zum Thema machen – und dies mit denjenigen phänomenologischen Mitteln, die die Phänomenologie selber erarbeiten half und die sie bereitgestellt hat.

In den folgenden Überlegungen soll versucht werden, ein Stück weit in diese Dimension des Fragens vorzudringen. Es liegt darin zugleich eine ganz bestimmte Beschränkung unserer Absicht. Sie soll lediglich entfaltet werden von Husserls eigenen Ansätzen her und ausschließlich im Rahmen seiner Analysen. Wir klammern also jede transzendente Stellungnahme zu Husserls Reduktionsmethode hier aus. Fragen ihrer mehr oder minder verborgenen metaphysischen Voraussetzungen, Fragen ihrer Motivierung, ihrer Rechtfertigung sollen als kritische Fragen vorsätzlich nicht in diese Betrachtung gehören. Es soll lediglich einmal systematisch der Frage nachgegangen werden, was diese Epoché eigentlich ist – und dies im Wege eines Versuchs, für sie selber eine Art *phänomenologischer Analyse* zu geben und zu untersuchen, was diese Epoché leistet, und zwar auf den verschiedenen Stufen der phänomenologischen, der transzendentalen und der thematischen Epoché. Ferner soll herausgestellt werden, wie die Reduktionen, die die Epoché ermöglicht, näherhin aussehen.

II

1. Die Selbstanwendung phänomenologischer Aktanalyse auf die Akte des Reduzierens selbst steht vor erheblichen Schwierigkeiten. Sie beginnen bereits, wenn wir von Reduktions '*akten*' sprechen. Handelt es sich denn hier tatsächlich um Akte, um Bewußtseinsweisen also von der Art, wie sie Husserl auch unter dem Titel intentionaler Erlebnisse faßt und die ihr Charakteristikum darin haben, auf etwas gegenständlich bezogen zu sein?

Wenn dem so ist, dann wird für die Epoché die Untersuchung ihrer noetisch-noematischen Struktur dringlich. Welches ist dann ihr Noema, und wie sehen die reduzierenden Noesen phänomenologisch aus? Kommt am Ende die Epoché unter den von Husserl analysierten Akten bereits selber vor? – Nur an einer frühen Stelle macht Husserl eine Andeutung in dieser Richtung. Wo es in den *Ideen* I um die Neutralitätsmodifikation doxischer Thesen geht, heißt es, daß diese der phänomenologischen Einklammerung 'nahe verwandt' sei.[3] Diese gehört also zunächst, wenn sie überhaupt einer Aktklasse zugerechnet werden darf, zu denjenigen Akten, die doxischer Art sind.

Damit ist zur phänomenologischen Charakterisierung der Epoché noch nicht viel gewonnen. Die Problemlage verschärft sich, wenn noch berücksichtigt wird, daß sich für Husserls Reduktionsverfahren im Grunde keine zutreffende Beschreibung findet und daß sich eine solche auch nicht finden lassen kann. Denn zwangsläufig müssen alle verfügbaren Ausdrücke in ihrem mundanen Wortsinn sich einer Methode versagen, die aus der Weltlichkeit hinausführen soll in eine transzendentale Dimension. Nichtsdestoweniger aber bleibt die verbale Mitteilung und Anweisung unumgehbar; und diese Aporie mag es bedingen, daß Husserl sich in beschreibender Absicht zu einer Vielfalt von Ausdrucksweisen genötigt sah und sie durch sein ganzes Werk hindurch auch beibehielt.

Dabei ist allerdings auffällig und für das Verständnis sogar erschwerend, daß die einzelnen Husserlschen Ausdrücke kaum als Synonyma angesehen werden können und sich in ihrer natürlichen Wortbedeutung keineswegs decken. (Die erste methodische Orientierung an Descartes' Zweifel, von Husserl später ausdrücklich nur als methodischen Behelf für den Einstieg in das Reduktionsverfahren bezeichnet, hatte ja – zumal sich die Motivierung der Reduktion in ihrer frühesten literarischen Form nicht scharf genug vom Traumargument des Descartes abhob – manchen

Mißdeutungen der Epoché Vorschub geleistet. Und wenn Husserl seine ersten Abgrenzungen gegen Descartes' Zweifel mit den Worten der 'Ausschaltung', der 'Einklammerung' belegte und diese auch durch ihren späteren wiederholten Gebrauch zu Termini fixierte, so waren auch sie eher geeignet, das grundsätzlich Neuartige der phänomenologischen Reduktion zu verdunkeln als zu erhellen.)

Es kann sich also gar nicht darum handeln, eine phänomenologische Analyse der Epoché mit der Analyse der Worte, die Husserl für sie gewählt hat, zu beginnen. Sie alle können nur die Funktion einer Andeutung haben – besser einer Anweisung: für ihren Vollzug. Was die Epoché ist, läßt sich einsehen nur im Wege ihres Vollzugs. Denn sie ist ja nichts außerhalb dieses Vollzugs, der, wenn er beständig und fortdauernd geübt wird, in seiner Habitualität dann 'phänomenologische Einstellung' heißt.

Wie wird diese phänomenologische Einstellung nach Husserl erreicht?

In natürlicher, unreflektierter Weltbezogenheit bin ich den Dingen schlicht hingegeben. Ich intendiere sie, aber ich bin darin nicht diesen intendierend zugewandt, sondern bin an das Intendierte verloren. Erst die Reflexion auf mein Intendieren läßt mich dessen inne werden, *daß* ich etwas intendiere. Diese erste Selbstreflexion ist eine 'natürliche' und hat ihr Kriterium darin, daß ich in ihr auch den Erfahrungsglauben der schlichten Intention teile; daß mir das Etwas, daß ich als ein Daseiendes, wirklich Seiendes intendiere, auch in der Reflexion noch weiterhin als dieses Daseiende, wirklich Seiende 'gilt'. Aber ich muß nicht als reflektierendes Ich immer mitglaubendes Ich sein. Ich kann mich in einer Weise reflektierend verhalten, daß ich mir diesen natürlichen Mitglauben versage – und dies in einer zweiten Reflexionsstufe, die auf mein Intendieren und sein Glauben geht. In ihr wird dieses Intendieren und sein Glauben nur noch betrachtet, ohne daß jedoch dieser Glaube mitvollzogen wird. Was ich damit vollziehe, – und zwar hier an einem einzelnen Akt –, ist eben nichts anderes als die Epoché, die 'Ausschaltung' des Seinsglaubens. Für mich, insofern ich sie ausübe, hat Husserl die nicht unmißverständliche Metapher des 'uninteressierten Zuschauers' gebraucht. Lieber möchte ich mich den 'unbeteiligten' Zuschauer nennen: Unbeteiligt bin ich auf dieser Reflexionsstufe am *Vollzug* des Seinsglaubens – aber ich inhibiere ihn nur deshalb, weil er mich *als* Vollzug meines naiven Ich in seinem Fungieren gerade interessiert.

Was wird mit dieser '*phänomenologischen*' *Epoché* erreicht, und wozu dient sie?

Die Beantwortung dieser Frage verlangt, daß ich mich in einer weiteren Reflexion nunmehr zum Zuschauer meiner selbst als phänomenologischen Zuschauers machen muß. Da zeigt sich, daß es den Sinn dieser Epoché erheblich verkürzen hieße, wollte ich ihn nur darin erblicken, daß mir die Epoché lediglich meinen schlichten Erfahrungsglauben in Sicht bringt. Schon der sprachliche Ausdruck für das Gesichtete verrät, daß die Leistung der Epoché früher anzusetzen ist. Denn wenn ich bereits meine natürliche Selbstreflexion dahingehend schilderte, daß sie den Erfahrungs*glauben* meiner schlichten Intention 'teilt', wenn ich sagte, daß ich als phänomenologisches Ich diesen Glauben '*in Geltung habe*', dann verrät diese Ausdrucksweise bereits ihre Abkunft aus meinem Aspekt als jener Zuschauer, der die Epoché bereits vollzogen *hat*.

Denn in natürlicher Selbstreflexion habe ich keinerlei Geltungsthematik. Zu ihr gehört nichts als die Intentionsrichtung auf den schlichten Aktvollzug, nicht aber werde ich mir darin auch schon des Seinscharakters des schlicht intendierten Etwas bewußt. Noch weniger habe ich darin ein positionales Bewußtsein von irgendeinem Mitvollzug doxischer Akte. Indem ich als Zuschauer vom '*Gelten*' eines Seinsglaubens spreche, genauer: indem ich mich in der Thematisierung dieses Geltens als phänomenologischer Zuschauer allererst etabliere, hat eine Art der Reflexion eingesetzt, die noch phänomenologische Bestände des naiven Aktvollzuges ans Licht hebt, die in der natürlichen Reflexion verborgen waren. Während ich in natürlicher Selbstreflexion das intendierte Etwas im Wie seines Vermeintseins als Wahrgenommenes, Erinnertes, Phantasiertes erfasse und damit allerdings auch die korrelativ zugehörigen Bewußtseinsweisen des Wahrnehmens, Erinnerns, Phantasierens, bringt mir die Epoché noch weitere mitbestätigte, in der natürlichen Reflexion aber verdeckt gebliebene Aktweisen in den Blick, nämlich die doxisch-thetischen und ihre Modifikationen sowie, korrelativ zugehörig, die doxischen Gegenstandscharaktere des Wirklichseins, Zweifelhaftseins usw. am intendierten Etwas. Die Epoché stellt sich mir also dar als die Ermöglichung der reflexiven *Thematisierung dieser Seinscharaktere* und ihrer korrelativ zugehörigen *doxischen Modalitäten*, d.h., der Thematisierung der Seinscharaktere *als* vermeinter im Wie ihres doxischen Vermeintseins – und das heißt im Sinne Husserls: als Seinsphänomene.

Was diese Epoché in bislang skizzierter Gestalt hergibt, ist also die Möglichkeit einer differentiellen Analyse der Bewußtseinsakte nach all ihren noetisch-noematischen Beständen – einer Analyse jedoch, die nach Ansatz und Zielsetzung der phänomenologisch betriebenen Psychologie zuzuweisen ist, d.h. einer Wissenschaft, die das Bewußtsein im mundan verstandenen Sinne als Bewußtsein der psycho-physischen Realität Mensch nimmt.

2. Die 'phänomenologisch-psychologische' Epoché ist aber nach Husserl ohne eigentliche philosophische Prätention. Für die Reduktion als philosophische Methode verlangt Husserl, daß sie *radikal* gefaßt und daß sie *universal* vollzogen werde. Nicht nur sollen einzelne Seinsgeltungen in singulären Intentionen inhibiert werden, sondern der gesamte Weltglaube mit allen Horizonten, mit allen Implikationen. Dieser *universalen Epoché* aber fällt eine völlig neuartige Aufgabe zu. Was ihren Einsatz nach Husserl motiviert, ist nichts Geringeres als die Begründung der Philosophie als Erkenntnis aus absoluter Rechtfertigung, die als ihren eigenen Boden keinen anderen zulassen darf als den, den sie in apodiktischer Gewißheit sich selber gibt.

Was dieser Probe einzig standhält, sind meine Cogitationes, und zwar meine *reinen* Cogitationes, die sich mir darbieten, wenn ich jene universale Epoché übe, die mit der gesamten Welt auch mich selbst zur Aussschaltung bringt, mich selbst als weltliches Wesen, d.h. meinem Leib und alle mit ihm zusammenhängenden Bedingtheiten meines Seelenlebens. Denn auch mein Leib – so das Husserlsche Argument – ist, obschon in einem besonderen Gegebenheitsmodus, ein Erfahrungsding wie andere Raumdinglichkeit und hat wie sie nur teil am Sein der Welt als Kontingenz mit der prinzipiellen Möglichkeit des Nichtseins.

Es ist nun zunächst zu versuchen, diese universale Epoché näher zu analysieren. Noch einmal gilt es, jene Stufen der Reflexion, die eben vordeutend an einer Einzelreduktion skizziert wurden, nachzuvollziehen und uns zum Zuschauer unserer selbst zu machen. Das heißt also, ich inhibiere jetzt allen Weltglauben, aber ich inhibiere ihn sozusagen nicht gerade hin, sondern ich *betrachte* reflektierend mein Inhibieren. Wie sieht das phänomenologisch aus? Was geschieht dabei?

Offenbar geschieht dabei überhaupt nichts; scheint doch diese Epoché, diese Enthaltung vom Seinsglauben, vielmehr ein bloßes Unterlassen zu sein. 'Einklammerung des Weltglaubens', 'Enthaltung von jedem Seinsurteil', 'Absehen von der Welt', 'Die Welt der Geltung berauben' – an all diesen Husserlschen Ausdrücken fällt ihre privative Form auf. Sie ist bedingt durch die Art des Einstiegs in das Reduktionsverfahren; sie verdeckt jedoch die eigentümlich positive Funktion, die der Epoché an sich zukommt. Die einzige positive Wendung, die Husserl ihr gibt, findet sich an einer allerdings sehr frühen Stelle. Dort heißt es, die Epoché sei "eine eigenartige Bewußtseinsweise, die zur ursprünglichen schlichten Thesis hinzutritt".[4]

Das gilt es genauer zu fassen, um einem naheliegenden Mißverständnis zu wehren. Denn diese Epoché bedeutet, recht verstanden, nicht eine Glaubensenthaltung, die einem schon als Seinsglauben *gewußten* Glauben widerfährt, so etwa, als sähe ich mich zunächst im Weltglauben, um diesen dann anschließend außer Vollzug zu setzen. Vielmehr bin ich, sobald ich mich im Weltglauben *weiß*, im Grunde schon aufgebrochen aus ihm, und die Epoché dient mir allererst dazu, mich dieses Aufbruchs zu vergewissern.[5] Die Epoché ist damit nicht einfach privativ die bloße Enthaltung dieses Glaubens, sondern sie ist von ihr selbst her etwas Positives – : Sie ist erst die *Aufdeckung dieses Glaubens als Glaube*. Denn mein *Urteil*, daß die Welt ist, gehört ja nicht selbst schon 'in' den naiven Glaubensvollzug, sondern wird erst 'aufgrund' seiner in reflektierender Einstellung möglich. Die universale Epoché macht also den Weltglauben als solchen allererst phänomenologisch *zugänglich*; sie bringt ihn reflektiv in den gewahrenden Blick, und er muß in ihm dann 'dahingestellt' bleiben, um als solcher im Wie seines Geltens – und das heißt für Husserl, als '*Phänomen*' – Thema der Phänomenologie werden zu können.

So stellt sich die universale Epoché also nicht einfach als eine Art doxischer Thesis mit umgekehrtem Vorzeichen dar, die einmal Gesetztes einfach zurücknimmt. Sie bietet sich noetisch – wenn wir in Analogie zu sonstigen Akten so sagen dürfen – als eine eigenartige Weise der *Bewußtwerdung doxischer Erlebnisse* meines naiven Vollzugsich, dem selber sie im Vollzug verdeckt bleiben. Das Was dieser Bewußtwerdung aber, ihr noematisches Korrelat also, ist dieses Erlebnis der Urdoxa, des Weltglaubens, der sich, voll entfaltet, dann für Husserl als ein ganzes System doxischer Thesen bietet und die als Akte ihrerseits ihre schlichte noetisch-noematische Struktur besitzen.

Es läßt sich also für die Epoché wohl in der Tat das Husserlsche Begriffspaar von Noesis und Noema verwenden. (Daß hier das noematische Korrelat der Epoché, kurz gefaßt als 'Weltglaube', selber dem Erlebnisstrom immanent ist, darf nicht befremden, da wir es bei der Epoché mit einer Aktleistung höherer Stufe zu tun haben. Deren Richtungsbestimmtheit auf Erlebnisweisen ist aber das phänomenologische Kriterium aller Reflexion überhaupt.) Nur ist wohl zu beachten, daß dieses Noema der Epoché eben seinerseits Akt resp. Aktsystem ist und damit seinerseits noetisch-*noematische* Struktur hat.

Diese Korrelatstruktur des reduzierenden Noemas ist zu berücksichtigen, wenn die Frage nach Sinn und Leistung der universalen Epoché gestellt wird. Übersieht man sie, so besteht die Gefahr, daß die Epoché mißverständlich ausgelegt wird, als sei der phänomenologische Zuschauer, der sie übt, nur noch den subjektiven Erlebnisbeständen zugewandt und am Sein der Welt gar nicht mehr interessiert. Es soll aber ja die universale Epoché nicht deshalb auf mein strömendes Aktleben führen, weil ich die Welt verabschieden will als etwas von ihm zu Trennendes, sondern weil mein Aktleben die Welt und alles Sein intentional in sich beschlossen hält. *Die Epoché gibt das Sein der Welt nicht preis, sondern sie behält es gerade ein*; sie macht es allererst ansichtig im Wie seines Als-seiend-Vermeintseins und eröffnet so die Möglichkeit, den Bezug der Subjektivität zum Sein durchsichtig zu machen und auf seinen absoluten Grund zu bringen.

Freilich ist für Husserl jede Seinsfrage nur zu stellen *innerhalb* der Subjektivität, die hier als 'transzendentale' nur erst vordeutend und prohibitiv benannt werden kann. In ihr verläuft nach Husserl alle Begründung und Ausweisung von Sein; und der Sinn, der dem transzendenten Sein zukommt, erhellt sich ganz und gar aus der Sinngebung der transzendentalen Subjektivität. Man kann diese Auffassung Husserls als 'idealistisch' abtun; aber man kann Husserl schlechterdings nicht vorwerfen, die Seinsfrage nicht gestellt zu haben. Und der oft gegen Husserl vorgebrachte Einwand, wie denn eine Philosophie den Sinn von Sein klären könne oder dies überhaupt nur könne unternehmen wollen, die das Sein 'ausgeschaltet' habe – diesen Einwand erheben hieße eben, die Epoché verkennen in ihrer eigenen noetisch-noematischen Struktur. Sie ist nicht einfach Abkehr von Sein, sondern ist *Zugang zum Erleben von Sein.* Indem sie thematisiert das In-Geltung-Haben allen Seins, macht sie offenbar die spezifischen Weisen des Bezugs zwischen Subjekt und Sein. "Man muß" – so beschließt Husserl seine Pariser Vorträge – "man muß erst die Welt durch Epoché verlieren, um sie in universaler Selbstbesinnung wiederzugewinnen"[6] – wiedergewinnen nämlich aus der letztbegründeten Einsicht, daß sie eine aus der transzendentalen Subjektivität geleistete und geschöpfte Welt ist.

Nun ist mit den letzten Andeutungen allerdings eine wesentliche Grenze innerhalb der Husserlschen Phänomenologie überschritten, die auch und gerade eine Leistungsgrenze der Epoché ist. Wohl nicht zufällig nennt Husserl die transzendentale Reduktion immer wieder eine Methode des *Zugangs* zur transzendentalen Sphäre: Des Zugangs – aber nicht ihrer Erforschung. Die Epoché bedeutet nur meinen Abschied von der Weltkindschaft, meinen Entschluß, der Weltverlorenheit zu entkommen. Aber sie gibt selber keine Wegweisung an die Hand, wie ich mein reines transzendentales Leben beginnen, wie ich es aus letztverantwortlicher Einsicht führen soll. Sie *eröffnet* mir das Feld meiner Cogitationes, aber sie *erschließt* mir dieses Feld nicht in seinen letzten befragbaren Gründen. Sie bietet mir mein Aktleben lediglich als fortwährend strömendes mit seinen intentionalen Korrelaten, mit dem, was sich in ihm konstituiert – 'konstituiert' in einem ersten und vorläufigen Sinne des Sichdarstellens, Darbietens von etwas. Aus einem tieferen Aspekt der Husserlschen Phänomenologie müßte es allerdings angemessener heißen: Mit dem, was sich in ihm immer schon konstituiert *hat.*

Denn die Epoché ist zwar *für mich*, in der Situation des Einstiegs, eine Zugangsweise zur transzendentalen Sphäre, aber sie ist *an und für sich* eine Haltung aus dem Perfekt; sie liefert *quasi* transzendentale Fertigprodukte. Doch gibt sie keinen Einblick in ihr Werden. Von der Systematik des Forschens her gesehen heißt das, daß die Epoché als Methode vor der eigentlichen Intentionalanalyse als Konstitutionsforschung endet. Diese folgt anderen methodischen Prinzipien, die hier nicht zu diskutieren sind – Prinzipien, die es ermöglichen, tiefer in dieses immerfort strömende Aktleben einzudringen, es in seiner Fertigstruktur aufzubrechen, um in seiner neuartigen transzendentalen Erfahrung sichtbar zu machen diese Akte als konstituierende, das heißt jetzt als *leistende, kreative Akte* in verschiedenen Schichtungen und Stufungen. Was die Epoché ansichtig werden läßt, sind zwar nicht andere Akte als 'leistende', aber sie gibt sie noch in der Verdecktheit ihres Leistungscharakters. Erst die weitere konstitutive Analyse führt mich auf mein Aktleben als 'letztlich leistendes Leben' und damit in eins zu dem, was die Welt als eine mir seiend geltende eigentlich ist –: Eine in meinen thetischen Leistungen immerfort einstimmig gesetzte und in fortlaufenden Synthesen der Bewährung als wirklich seiend ausgewiesene Welt. Erst auf dieser Stufe der Untersuchung klärt sich auch der präzisere Sinn des Transzendentalen bei Husserl. Es ist dasjenige, das alles transzendente Sein in sich konstituiert, aus dem alles transzendente Sein seinen Seinssinn schöpft. Und dieses Transzendente liegt nicht außerhalb der transzendentalen Subjektivität, sondern ist in der Weise irrealen Beschlossenseins selber zur transzendentalen Immanenz gehörig.

Augenscheinlich gelangt man hier allerdings in einige Bedrängnis. Wie kann sich innerhalb der transzendentalen Subjektivität echte Transzendenz bekunden? Ist diese Subjektivität absolutes und in sich geschlossenes Sein, wie Husserl wiederholt betont[7], ist für sie die Annahme eines Außerhalb jedweden Sinnes schlechterdings sinnwidrig, wie ist dann Transzendenz, Objektivität, Ansichsein überhaupt noch zu denken möglich?

3. Auch für diese Frage ist es ergiebiger zu sehen, was Husserl tut, als was er sagt. Er tut Entscheidendes für dieses Problem in der 5. Cartesianischen Meditation.[8] Sie stellt inhaltlich einen Entwurf zur Lösung der Intersubjektivitätsfrage dar, bietet jedoch zugleich eine methodisch sehr konsequente, aber auch sehr eigenartige Durchführung der transzendentalen Reduktion.

Überblicke ich das Feld, das mir die universale Epoché für die transzendentale Erfahrung eröffnet hat, so stelle ich zunächst fest, daß es *mein* Feld ist – strömendes und weiterhin leistendes Leben meines transzendentalen Ego. Die transzendentale Epoché führt so scheinbar zwangsläufig zum transzendentalen Solipsismus. Die anderen Subjekte sind in meinem transzendentalen Erfahrungsfeld nur phänomenale Vorfindlichkeiten mit Seinsanspruch, den möglicherweise ablehnen zu müssen überdies keinerlei Widersinn ergäbe. Ein konsequenter Solipsismus müßte aber gerade vor der Aufgabe versagen, die Husserl sich gestellt hat: Die Objektivität der Welt wird aus meinen alleinigen konstituiven Leistungen nicht verständlich. Den Sinn des Wirklichseins kann ich in mir allein ausweisen, den der Objektivität nicht. Denn Objektivität enthält, als Wirklichkeit für jedermann, für beliebige andere, ein im Sinn Überschüssiges, eben dieses Sinnmoment von anderem Subjektsein. Es gilt also zu klären und aus meinen konstitutiven Leistungen zu transzendentalem Verständnis zu bringen, was den Sinn des Seins von anderen ausmacht.

Gegeben ist mir der Andere als Leib. Leiblichkeit ist überhaupt die Gegebenheitsform für fremde Subjektivität. Aber diese ist mir in ihrer Leiblichkeit nicht originär gegeben, sondern 'zeigt' sich in ihr nur 'an'. Was aber ist der phänomenale Sinn solchen Anzeigens? Auch dieser Sinn muß sich mir nach der transzendentalen Epoché darstellen als ein von mir geleisteter, aus meinen eigenen konstitutiven Leistungen gestifteter Sinn. Husserl bezeichnet diese Leistung als 'Appräsentation'. Wo liegen ihre transzendentalen Bedingungen?

Husserl geht dieses Problem an, indem er es – scheinbar – zunächst beiseiteschiebt. "Aus methodischen Gründen", so heißt es dort, rechtfertige sich eine zweite Epoché, die sogenannte "*thematische Epoché*". Sie stellt noch *innerhalb* der universalen Epoché einen neuen Reduktionsschritt dar.

Mit dieser thematischen Epoché sollen ausgeblendet werden alle Bestimmungen der phänomenalen Welt, die in ihrem Sinn auf andere Subjekte verweisen. Es entfallen also damit in allen gesetzten Seinsgeltungen solche Sinnkomponenten, die ich sozusagen nicht allein konstituieren kann. Noetisch gesprochen: Ich *durchstreiche* jetzt vorübergehend alle durch fremde Subjektivitäten mitgeleistete Sinnkonstitution. Wir wollen hier mit Husserl annehmen, daß das möglich sei. Damit spaltet sich mein transzendentales Erfahrungsfeld in eine Sphäre des Fremden und eine Sphäre des Eigenheitlichen. Dieses Eigenheitliche erstreckt sich auf die apodiktische Evidenz des Ich-bin in seiner universalen Strukturform und in all seinen konstituierten Einheiten. Damit verbleibt eine zusammenhängende Schicht der Welt, in der die anderen Subjekte reduziert sind auf raumdingliche Phänomene wie andere Dingphänomene. Wie komme ich dazu, einige solcher Dingphänomene als '*Leib*'phänomene aufzufassen?

Unter allen Raumdingen begegnet mir in meiner Primordialsphäre ein ausgezeichnetes: Mein eigenes Leibding. Mit ihm hat es in der Husserlschen Phänomenologie eine eigentümliche Bewandtnis. Die universale Epoché reduzierte meinen Leib auf ein – übrigens kontingentes – Faktum in meinem reinen, transzendentalen Bewußtsein, d.h. auf das Phänomen meines Leibes im Wie seines Erscheinens, dessen Seinssinn es in bestimmten Erfahrungen auszuweisen und zu bewähren gilt.

Nach der thematischen Epoché auf die Eigenheitssphäre gilt es nunmehr, diese Erfahrungsweise genauer zu analysieren. Dem ist hier im einzelnen nicht nachzugehen. Dazu sei hier nur das Wesentliche herausgehoben. Meine Leibeserfahrung ist in ihrer spezifischen Weise angezeigt als Empfindung. Eine besondere Rolle spielen im Getast die sogenannten Doppelempfindungen, vermöge deren sich so etwas wie ein 'Innen' und 'Außen' erfahrungsmäßig darbietet.

In dieser Innen-Außen-Erfahrung konstituiert sich nach Husserl der ursprüngliche Sinn von Transzendenz. Noch ist sie nicht die volle Transzendenz der objektiven Welt. Immerhin aber wird bereits in meiner Primordialsphäre eine transzendente Welt konstituierbar. Mit meinem Leib apperzipiere ich mich jetzt als etwas in der Welt, als Teil ihrer mit einem Außer-mir, mit ihm fasse ich mich auf als Mensch. Und nicht nur mich. Denn außer mir gibt es in meiner primordialen Erfahrung andere Leiber. Aber was heißt gerade dieses 'es gibt'? – Das Husserlsche Reduktionsverfahren erhält seine letzte methodische Rechtfertigung, wenn es dieses 'es gibt' der Anderen aufzuklären und transzendental verständlich zu machen vermag. Denn an ihm hängt wesentlich der Sinn von Sein als Ansichsein, Transzendentsein, Objektivsein – und damit letztlich von Wahrsein.

Husserls Antwort darauf lautet: Es gibt die Anderen 'aufgrund' meines Leibes. Das besagt näherhin, daß ihr Sinn als 'Leib' ihnen zuwächst kraft der Ähnlichkeit mit meinem Leib. Der andere Leib muß den Sinn seines Seins von meinem übernehmen. Solche Sinnüberschiebung erwirkt weiter, daß der andere Leib nicht bloß als solcher, sondern auch als Leib eines anderen Ich mit eigenem Seelenleben da ist. Auch dessen 'Anzeige' im anderen Leib ist mithin meine eigene appräsentierende Leistung, und zwar eine wohlmotivierte: Weil ich selbst mich als Leib erfahre mit eigenem Seelenleben, muß auch jeder andere Leib sich in gleicher Weise wissen, und dies nicht nur 'für' mich, sondern auch 'dank' meiner – und letzten Grundes dank meines Leibes.

Mit dieser Sinnstiftung des Anderen als des anderen Ich geschieht zugleich aber etwas in meiner Selbstapperzeption. Ich bin mit meiner Konstitution des Anderen erheblich rückbetroffen, insofern ich ihm als dem anderen Ich alle diejenigen Leistungen zuzudenken habe, die auch mein wahres Sein ausmachen. Der Andere ist nicht nur überhaupt Ego in der Welt, sondern auch konstituierendes Ego *für* die Welt, er ist transzendentales Ich, wie ich es bin. Mit der Konstitution des Anderen also apperzipiere ich mich erst in dem, was ich eigentlich bin – Glied einer Gemeinschaft transzendentaler Egos, die miteinander in den mannigfachen Verflechtungen implikativer Sinnstiftungen die eine Welt konstituieren. Ich bin nicht als *solus ipse* schon der ganze Grund der Welt, den zu finden ich mit der transzendentalen Epoché aufgebrochen war, sondern erst die transzendentale Wir-Gemeinschaft macht die volle transzendentale Subjektivität aus.

Damit scheint das Gespenst des Solipsismus gebannt; damit scheint zugleich die Sinnklärung von Objektivität und Transzendenz der Welt geleistet: Sie weist sich als solche aus in der Harmonisierbarkeit und Identifizierbarkeit der verschiedenen Weltbilder aller monadischen Egos; und sie hat in dieser beständigen Ausweisung in der transzendentalen Wir-Gemeinschaft ihren alleinigen, aber auch zureichenden Rechtsgrund.

III

Vergegenwärtigen wir uns abschließend kurz noch einmal das Eigentümliche dieser thematischen Epoché und vergleichen sie mit der transzendentalen. Die thematische Epoché setzte an demjenigen Phänomenbestand an, den mir die universale Epoché geliefert hatte. Während diese letztere die Reduktion der Welt auf das transzendentale Weltphänomen leistet, bedeutet nun die thematische Epoché eine ganz bestimmte methodische Scheidung innerhalb dieses Weltphänomens, nämlich die Scheidung von Fremdsphäre und Primordialsphäre. Der Sinn dieser thematischen Epoché ist also ein wesentlich anderer als der der transzendentalen Epoché. Er kommt dem Wortsinn der 'Ausschaltung', der 'Einklammerung' erheblich näher als die transzendentale Reduktion. Hier wird in der Tat etwas zunächst methodisch durchstrichen, nämlich alles Fremdsubjektive, soweit es das eigene Fungieren und Konstituieren betrifft. Aber auch für die thematische Epoché gilt, daß sie nicht *nur* diese Durchstreichung bedeutet, sondern daß sich zugleich in und mit solcher Durchstreichung für mich diejenige Bewußtseinsstellung ergibt, in der mir außerhalb meines transzendentalen Erfahrungsfeldes das Phänomen der Anderen, Fremden überhaupt er *als Phänomen zum Thema* werden kann.

Das ist etwas sehr Bemerkenswertes. Denn woraus bezieht diese Reduktion auf die Primordialsphäre ihren Sinn, und worin gipfelt ihre Leistung?

Offensichtlich trägt sie dem Umstand Rechnung, daß die 'andere' Subjektivität von vornherein erscheint als Teil der Welt, besser, als Stück meines transzendentalen Weltphänomens. Das ist nun so lange nichts Absonderliches, als man mit Husserl daran festhält, daß dieses Weltphänomen doch das *Insgesamt aller Kontingenzen* bildet, zu denen dann, wie der eigene Leib, so auch die fremden Leiber gehören. Weil – zufällig – zu dem Anderen wie zu mir ein Leib gehört, muß er sich eben als dieses kontingente Faktum im Rahmen der Welt zeigen und in seinem Seinssinn klären lassen. Dieser klärt sich nach dem Voraufgegangenen aufgrund einer ganz bestimmten Erfahrungskontinuität mit eigenem Bewährungsstil, darin sich der Leib als seiender ausweist. Nichts anderes bedeutet die Rede Husserls, daß ich, als transzendentales Ego, mich als Mensch in der Welt 'konstituiere'.

Nun ist aber, wie mir die thematische Epoché zeigt, diese verweltlichende Selbstapperzeption doch ganz offenkundig nicht nur eine transzendentale Leistung unter anderen. Vielmehr ist sie dadurch eigentümlich ausgezeichnet, daß das in ihr Geleistete, nämlich der Sinn meines Leibseins, seinerseits zum Grund eines Leistens wird, nämlich der Konstitution fremden Subjektseins. Anders gesagt, was mir die universale Epoché als eigenes Leibphänomen, als eigenes Weltstück gab, das hat mir die thematische Epoché dahingehend zur Auslegung gebracht, daß es der Seinssinn dieses merkwürdigen 'Phänomens' Leib ist, nun seinerseits Grund, und zwar *leistender* Grund zu sein für andere 'Phänomene' in der Welt. Denn die Fremdsubjektivität erhält ihren Sinn erst durch Einlegen meines Eigentlichen in die anderen Körper resp. Leiber.

Der Sinn meines Leibes ist also einerseits der, daß er transzendental geleisteter, gestifteter Sinn eines Stücks Welt ist – zum anderen aber auch der, daß er seinerseits *Bedingung* für Sinnstiftung von Welthaftem ist – und überdies gerade für solche Sinneinheiten der Welt, denen ich ihrerseits konstitutive Funktion für mein volles, konkretes Sein zuzudenken habe.

Das aber besagt ein doppeltes. Es besagt erstens, daß die Bedingung, als die mein Leib (und somit jeder Leib) in der Konstitution der Welt fungiert, eine *notwendige* Bedingung ist. Denn erst mit der Vielzahl der Leiber erhält die transzendentale Subjektivität Pluralstruktur; aber sie erhält sie auch *nur* durch die Pluralität der Leiber. Erst mit ihr also wird so etwas wie Subjekt-Gemeinschaft, Intersubjektivität überhaupt möglich und daher mit ihr auch erst die Konstitution von Objektivität. Demnach gehört aber Leiblichkeit wesensnotwendig zur vollen konstituierenden Subjektivität.

Schließlich hat sich zweitens ergeben: Der Sinn des Seins der Welt erschöpft sich nicht darin, ausschließlich in mir ihren konstitutiven Ursprung zu haben. Ihr Seinssinn ist vielmehr so zu denken, daß die Welt wiederum konstitutiver Grund und Ursprung ist für das, was ich bin – und so für jedes Ego. Subjekt und Sein, Ich und Welt stehen somit nicht in einem Verhältnis einseitiger Abhängigkeit, sondern bilden eine synthetische Einheit zweier Glieder, die nur in wechselseitiger Implikation bestimmbar und auslegbar sind.

Methodologisch betrachtet nimmt sich dieses Resultat einigermaßen wunderlich aus, und eine phänomenologische Theorie der Epoché, eine Betrachtung also, die nach Husserl Sinn und Leistung der Epoché im ganzen zu überdenken hat, gerät hier wohl in einige Verlegenheit.

War einerseits die *transzendentale Epoché* eigens deshalb als universale gefordert, damit ich mich von aller weltlichen, kontingenten Bedingtheit frei, rein als mich selbst fasse, extramundan in der apodiktischen Evidenz meines reinen transzendentalen Lebens, so hat nun die *thematische Epoché*, einsetzend innerhalb der universalen, mich zu einer apodiktischen Einsicht geführt, daß ich mich als diese Sinn und Sein leistende Subjektivität gerade nur begreifen kann, wenn *Weltlichkeit selbst* schon *notwendig* zu meinem Wesen als konstituierendes Ego gehört.

War ferner die transzendentale Epoché eben dadurch motiviert, daß sie auf ein Feld absoluter Ursprünge, auf einen letzten irrelativen Seinsboden der Welt in der reinen, transzendentalen Subjektivität führen sollte, aus der die Welt ihren ganzen Seinssinn schöpft, so hatte hinwiederum gerade die thematische Epoché bereits ihre Motivation gerade darin, daß alle Subjektivität nur erscheint auf dem Boden der Welt und hat so gerade zu dem Aufweis eines streng wechselseitigen Leistungsverhältnisses von Subjekt und Welt geführt.

Es bleibt abschließend die Frage, ob nicht damit die thematische Epoché, so wie Husserl sie darstellt, in der konsequenten Verfolgung ihres eigenen Ansatzes schließlich zur phänomenologischen Selbstaufhebung der universalen Epoché treibt.

Wir sind uns bewußt, daß solche Fragen das Herzstück der Husserlschen Phänomenologie angehen. Aber es könnte sich bei tieferem Nachdenken erweisen, daß, selbst wenn diese Fragen sämtlich bejahend ausfallen müßten, die universale Epoché damit nicht schon sogleich als ein verfehlter Schritt dieser Phänomenologie angesehen werden müßte, sondern als eine methodische Maßnahme, die zwar vorläufig und überholungsbedürftig ist, deren Vorläufigkeit aber transzendental zu durchschauen sie selber als letzte methodische Konsequenz aus sich hervortreibt.

Technische Universität Braunschweig

ANMERKUNGEN

1 *Ideen*, I, *Hua*, III, § 63; *Ideen*, III, *Hua*, V, S. 93, 141f.; vor allem *Erste Philosophie*, 2. Teil, *Hua*, VIII, 40–43. Vorlesung u. Beilage II, S. 313; s.a. *Krisis*, *Hua*, VI, § 44. Die Werke Husserls werden in den üblichen Abkürzungen des Titels nach der Gesamtausgabe der *Husserliana*, Den Haag 1950ff. (hier als *Hua*) zitiert.

2 a.a.o. *Hua*, VIII, Beilage II, S. 313.

3 *Ideen*, I, *Hua*, III, § 109.

4 *Ideen*, I, *Hua*, III, § 63.

5 Deutlich ist diese Bedeutung der Epoché bereits herausgestellt worden von E. Fink in: 'Die Philosophie Edmund Husserls in der gegenwärtigen Kritik', *Kant-Studien* **38** (1933) 319–383.

6 *Cart. Med.*, *Hua*, I, S. 39.

7 Vergl. u.a. *Ideen*, III, *Hua*, V. S. 153; *Form. u. Transz. Logik*, Halle 1929, § 94; *Krisis*, *Hua*, VI, §§ 41, 49.

8 Hierzu u. zum folgenden *Cart. Med.*, *Hua*, I, §§ 42ff.

K. KUYPERS

DIE WISSENSCHAFTEN VOM MENSCHEN UND HUSSERLS THEORIE VON ZWEI EINSTELLUNGEN

Das Verhältnis, in dem Husserl zu Dilthey steht, hat sich in der späteren Periode seines Lebens wesentlich geändert. Bekämpfte er im Anfang in seiner Logosabhandlung in einer scharfen Auseinandersetzung Diltheys Historismus, so anerkannte er nachher vollauf die Bedeutung der Historizität in der Kultur und im menschlichen Dasein. Auch kam er zu einer völligen Anerkennung von Diltheys Theorie der Geisteswissenschaften, jedoch mit bestimmten Einschränkungen und mit kritischen Bemerkungen über das Fehlen scharfer Formulierungen und methodisch sicherer Lösungen.

Mit der Anerkennung von Diltheys Verdiensten ist die Erkenntnis verbunden, dass der Aufbau einer allgemeinen Wissenschaft des Geistes und das Erwerben von Einsicht in das Verhältnis, in dem die Phänomenologie zu dieser Wissenschaft steht, nichts Geringeres als einen neuen Weg in die Phänomenologie bedeuten würde, der sogar – wie er meint – viel besser und weiterreichender als der von der Psychologie sein würde.

Da die Phänomenologie ihr Entstehen der deskriptiven Psychologie von Brentano verdankt und Husserl trotz allem diesem Ursprung immer treu geblieben ist, ist die Anerkennung von Diltheys Bestreben um so bemerkenswerter. Es erhebt sich jedoch die Frage, was dieser neue Weg zu bedeuten hat.

Dieser hängt ja engstens mit einer Auffassung zusammen, die auf einer typisch deutschen Tradition beruht, einem Dualismus in der Wissenschaftstheorie, der Unterscheidung nämlich zwischen Natur- und Geisteswissenschaften, welche ihren Ursprung bei Hegel hat, aber besonders durch Dilthey – jedoch nicht ohne einiges Zögern seinerseits gerade hinsichtlich der Benennung – zu einer festen Terminologie geführt hat, welche sich übrigens schwer in anderen Sprachen wiedergeben lässt. Zwar gehört es zur französischen und angelsächsischen Tradition, dass man einen Unterschied macht zwischen *science* einerseits und *arts* und *humanities* andererseits. Dieser schafft insoweit keinen Dualismus in der Wissenschaftstheorie als die *arts* und *humanities* einfach kraft des Be-

Tymieniecka (ed.), Analecta Husserliana, Vol. I, 186–196. *All Rights Reserved*
Copyright © 1970 *by D. Reidel Publishing Company, Dordrecht-Holland*

griffes von Wissenschaft, den man hantiert, ausgeschlossen werden. Zwar bestrebt man sich sowohl in der Wissenschaftstheorie als in der Praxis der wissenschaftlichen Betätigung, auch die Psychologie und die sozialen Wissenschaften unter dem Begriff von *science* unterzubringen und den dazu gültigen Normen und Forderungen entsprechen zu lassen. In der deutschen Wissenschaftstheorie macht man aber einen prinzipiellen Unterschied zwischen Naturwissenschaften einerseits und allen jenen Wissenschaften, welche menschliches Bewusstsein, Benehmen und Leistungen zum Thema haben, andererseits. Diese fasst man dann zusammen unter dem Ausdruck 'Geisteswissenschaften'. Gegen den Dualismus erhebt der *Unity of Science-Movement* am stärksten Widerspruch. Diese findet seine schärfste Formulierung in Carnaps Physikalismus.

Die Phänomenologie im allgemeinen und namentlich Husserl hat sich für den Dualismus eingesetzt. Man kann sogar sagen, dass Husserl eine philosophisch stärkere und theoretisch verbesserte Ausgabe von Diltheys Theorie der Geisteswissenschaft hat geben wollen, und zwar indem er die Phänomenologie als universale Wissenschaft des Geistes oder allgemeiner der Subjektivität erhebt. Neben der Anerkennung der Bedeutung der Historizität als eines wesentlichen Momentes steht also auch die Übernahme und Fortsetzung der Aufgabe, dem, was in der deutschen Wissenschaftstheorie Geisteswissenschaften genannt zu werden pflegt, ein philosophisches Fundament zu geben. Dabei ist die Kritik Husserls an Dilthey (z.B. die Art und Weise, wie Dilthey das Fundament mittels einer Unterscheidung zwischen Beschreiben und Erklären gelegt hat, welche Husserl mit Recht ablehnt) nebensächlicher Bedeutung. Das Wichtigste scheint uns, dass mit dieser Aneignung eine nicht unwichtige Verschiebung in der Auffassung und Zielsetzung der Phänomenologie stattfindet. Diese Verschiebung besteht in der Aneignung jener Idee einer allgemeinen Wissenschaft des Geistes als eines neuen, besseren und weiterreichenden Weges in die Phänomenologie. Erst mittels der Theorie der Reduktionen hat sich die Idee der Phänomenologie, entstanden aus der Psychologie von Brentano, zu einer universalen, und zwar transzendentalen Philosophie entwickeln können. Wir können hier nicht näher auf die Schwierigkeiten eingehen, welche dieses hervorrief und weisen nur darauf hin, dass Husserl, während er neben Frege den damaligen Psychologismus ablehnte und bekämpfte, dessen ungeachtet u.a. von Carnap den Vorwurf zu hören bekam, selbst den Psychologismus nie ganz überwunden zu haben.

In der neuen Auffassung der Phänomenologie, in der diese in einer universalen Wissenschaft des Geistes untergebracht wird, ist es zweckmässig, drei Momente zu unterscheiden. Erstens die anthropologische Unterscheidung zwischen Körper (Leib), Seele und Geist (Person), zweitens die Theorie von zwei verschiedenen Einstellungen, welche man annehmen kann, was zu zwei verschiedenen Arten von Gegenständen und Wissenschaften führt, und drittens den metaphysischen Hintergrund und Zusammenhang, in denen die Theorie der Geisteswissenschaften zu der Philosophie des Geistes oder des absoluten Bewusstseins oder Geistes steht.

Husserl spricht unumwunden in der Einleitung zur *Konstitution der geistigen Welt* (*Ideen*, II; *Husserliana*, IV, S. 172) aus, dass es sich hier um metaphysische und wissenschaftstheoretische Unterscheidungen handelt, insbesondere um die Unterscheidung zwischen Seele und Geist und dass die Aufklärung dieser Unterscheidungen unsere ganze Weltanschauung grundwesentlich bestimmt.

Sehr deutlich geht hieraus hervor, dass hier nicht nur rein wissenschaftstheoretische Gründe zur Geltung gebracht werden, sondern die uralte metaphysische Polarität von Natur und Geist das eigentliche Motiv und den Hintergrund der wissenschaftstheoretischen Kontroverse bildet. Die Einteilung in Natur- und Geisteswissenschaften wird schliesslich durch eine Philosophie des Geistes beherrscht. Dass diese besonders in der deutschen Wissenschaftstheorie solch eine grosse Rolle zu spielen angefangen hat und mehr oder weniger zu einer festen Tradition geworden ist, wird wohl dem Einfluss von Hegel zugeschrieben werden müssen. Dass bei Husserl die anthropologische Dreiteilung in Körper (Leib), Seele und Geist von neuem solch eine grosse Bedeutung bekommt im Gegensatz zu der angelsächsischen und cartesianischen Zweiteilung, darf man aber nicht einfach als eine Rückkehr zu Aristoteles und Hegel auffassen, obwohl die metaphysische Tradition in diesem Punkt bei Husserl sich unleugbar geltend macht. Die Dreiteilung beruht vielmehr auf einem Versuch, der Seele als Realität eine vermittelnde Rolle und Funktion zuzuweisen in dem polaren Gegensatz zwischen Natur (Körper) und Geist.

Obwohl die genannten drei Momente in Husserls Konstitutionstheorie aufs engste miteinander zusammenhängen und die kritische Wertbestimmung am Ende nur von diesem Zusammenhang aus geschehen kann, wollen wir uns hier beschränken auf Husserls Theorie der zwei Einstellungen: der naturalistischen und der personalistischen oder geisteswissenschaftlichen. Diese ist für die Hermeneutik und Wissenschaftstheorie der heutigen Zeit im Zusammenhang mit der wachsenden Tendenz, auch in den Wissenschaften vom Menschen zu Objektivierung und Formalisierung und schliesslich sogar zu Mathematisierung zu kommen, aktuell und von besonderer Bedeutung. Die Objektivierung wird als ein Weg gedacht, der nicht nur zu Exaktheit, dem Ideal aller Wissenschaft, sondern auch zur echten Wissenschaft und so zur Einheit aller wissenschaftlichen Methoden und Techniken hinführt. Dieses Fortschreiten der Objektivierung und Formalisierung wird als ein Fortschritt betrachtet und als eine Verdrängung weniger exakter Denkweisen und Methoden.

Die Stellung der Theorie der Geisteswissenschaften ist im allgemeinen defensiv und konservativ hinsichtlich der bestehenden Praxis bestimmter Wissenschaften, namentlich der historischen, während der Gegner, welchen wir in seiner extremen Form als Vorkämpfer des Physikalismus bezeichnen können, wohl oder nicht verbunden mit dem Namen Carnaps, in einer offensiven Lage ist und in seiner Forderung der Einheit der Wissenschaften eine mögliche zukünftige Entwicklung aller empirischen Wissenschaften nach dem Modell der Naturwissenschaft vorwegzunehmen meint.

Was versteht Husserl nun unter den zwei Einstellungen? Kurz zusammengefasst läuft dieses auf Folgendes hinaus: Husserl geht aus von einer natürlichen Einstellung der vorgegebenen Welt gegenüber, die immer als Umwelt erscheint und für jeden im Prinzip immer anders gegeben sein kann. Diese ist die Lebenswelt, das heisst die Welt, in der wir leben, wobei die Natur so wie sie uns erscheint aufgefasst wird. Aus dieser Lebenswelt entwickelt sich zweierlei wissenschaftliche Einstellung, die naturwissenschaftliche und die personalistische oder geisteswissenschaftliche. Die erstgenannte ist auf Erforschung der physischen Natur an sich und von Wahrheiten an sich gerichtet. Dieses geschieht durch Idealisierung, das heisst durch das Entwerfen von Modellen und Formalisierung mit Hilfe der Mathematik. Es ist aber auch möglich, dass man die ganze Welt als Natur auffasst. In diesem Falle erfährt der Naturbegriff eine Erweiterung, die mit dem übereinstimmt, was für die Griechen ursprünglich auch die *physis* war: Alles was da ist, die All-Natur, nachher beschränkt auf das, was Husserl die physische Natur nennt. Damit ist eine universale induktive Erfahrung Korrelat und der Anspruch einer universalen Naturwissenschaft, in der alle Erscheinungen ohne Ausnahme auf grund induktiver Erfahrung erforscht werden. Dieses enthält zugleich die Anerkennung des Rechtes unbeschränkter Anwendung des Prinzipes der Objektivierung und Formalisierung beziehungsweise Idealisierung auch hinsichtlich der geistigen Sphäre und der geistigen Subjektbeziehungen. Mit der Erweiterung des Naturbegriffes verbindet sich deshalb das, was Husserl das naturalistische Vorurteil nennt, nämlich die Meinung, dass Natur auch das wahre Sein des Subjektes sei. Der Naturalismus betrachtet den Menschen ausschliesslich nur als 'erfüllte Extension' und so die Welt überhaupt nur als eine 'erweiterte Natur'. In der entgegengesetzten Einstellung, der geisteswissenschaftlichen oder personalistischen, bilden gerade diese geistigen Subjektbeziehungen das Untersuchungsfeld. Hier bleibt der Mensch als Person gegeben, wie das auch in der natürlichen Einstellung der Fall ist, und auch die Natur, wie diese darin erscheint. Dadurch unterscheidet sich diese Art von Wissenschaften prinzipiell von den Naturwissenschaften. Der naturalistischen Einstellung entgeht, obzwar sie das All der Realitäten zu untersuchen meint, eben gerade die Lebenswelt der Personen und auch die meist subtile naturwissenschaftliche Theorie kann dieser nicht beikommen. Man kann nun kritisch die Frage stellen, welche Wissenschaften zu den Geisteswissenschaften gehören und sich in dieser Bezeichnung zusammenfassen lassen. Im allgemeinen besteht die Neigung, diese mit den Wissenschaften vom Menschen als Menschen zu identifizieren (menschliches Bewusstsein, Benehmen, Leistungen, Institutionen, Kulturgebilde). Sowohl bei Dilthey als bei Husserl ist besonders die Stellung der Psychologie problematisch. In Übereinstimmung mit den Auffassungen am Ende des 19. und am Anfang des 20. Jahrhunderts betrachtete auch Dilthey die Psychologie als Grundwissenschaft für alle Wissenschaften vom Menschen und erhob nun das Postulat einer Wissenschaft, welche dieser Forderung genügen könnte und als Grundlage für die Geisteswissenschaften, auch von Husserl als beschreibende, konkrete Kulturwissenschaften bezeichnet, geeignet sei. Das führte zur Anerkennung zweier Möglichkeiten und Richtungen der Untersuchung und Methode innerhalb der Psychologie: Einer naturwissenschaftlichen und einer geisteswissenschaftlichen. Die Anerkennung von zwei ganz verschiedenen Methoden und Richtungen innerhalb einer und derselben Wissenschaft ist an sich sehr merkwürdig. Sie enthält ja die Anerkennung nicht nur eines prinzipiellen Dualismus hinsichtlich der

Wissenschaften im allgemeinen, sondern auch innerhalb einer bestimmten Wissenschaft, in diesem Falle der Psychologie.

Man findet die Anerkennung des Rechtes beider Richtungen in Husserls Theorie der zwei Einstellungen in *Ideen*, II und besonders in den Beilagen, wo sie näher entwickelt wird, während eine Beilage (datiert: vor 1930) im Bande VI der *Husserliana* (*Die Krisis der europäischen Kultur*) die reifste Formulierung und eine sehr suggestive Perspektive bietet. Während Husserl die Phänomenologie selbst auf einer bestimmten Einstellung gründet und diese Idee in seiner Lehre der Reduktionen näher entwickelt, tut er Ähnliches auch hinsichtlich der Wissenschaften vom Menschen.

Von der Einstellung hängt ab, was als Gegenstand gegeben ist und welche Methode man darauf anwendet. Wendet man die naturalistische Einstellung an, dann werden Seele und Geist der Natur untergeordnet und als ein Teil davon betrachtet. Wendet man dagegen die geisteswissenschaftliche Einstellung an, dann ist der Mensch beziehungsweise Mitmensch als Person gegeben, dem die Welt als Umwelt erscheint und der handelt, denkt und wertet im Rahmen der natürlich gegebenen Welt. Die naturalistische Einstellung hat mit der Einstellung der Naturwissenschaft die Intention gemein, sich nicht zu richten auf die Erscheinungsweise, sondern auf das An-sich der Erscheinungen und deshalb auch auf Wahrheiten an sich.

Das Bemerkenswerte ist, dass Husserl im Prinzip das Recht beider Einstellungen anerkennt. Dieses trifft besonders zu für die Psychophysik, welche den Menschen als reale, doppelschichtige Einheit in der objektiven Natur zum Thema hat und die Abhängigkeitsbeziehungen von Seele und Geist in Beziehung zum Körper untersucht. Die Anerkennung geschieht aber mit einer gewissen Einschränkung. Der Geist kann zwar als abhängig von der Natur betrachtet und auch selbst naturalisiert werden, aber nur bis zu einem gewissen Grade, weil sonst das fehlen würde, was der Natur Sinn gibt (*Husserliana*, IV, S. 297). Die Natur ist ein Feld durchgängiger Relativitäten. In letzter Instanz sind diese alle jedoch relativ zu einem Absoluten, das alle Relativitäten trägt und das ist der Geist.

Ausserdem hat die theoretische Anerkennung des Rechtes einer universalen Welt- oder Naturwissenschaft, in der das Universum als in einer universalen Wissenschaft aus Axiomen deduzierbar nach dem Beispiel der Geometrie oder aus Hypothesen nach dem Beispiel der Physik gedacht wird, eine Kehrseite, welche die Bilanz zugunsten der Geisteswissenschaften zum Ausschlag bringt.

Mit der Idee der Fundierung der Geisteswissenschaften in einer allgemeinen Wissenschaft des Geistes verbindet Husserl ja die Idee der Phänomenologie selbst. Die Verteidigung der Methode der Geisteswissenschaften als auf konkrete Anschauung gerichtet mit der Welt, wie diese in der unmittelbaren Anschauung erscheint, der Lebenswelt, also der Welt, in der wir als Personen leben, als Ausgangspunkt der Geisteswissenschaften, beruht ja darauf, dass die Phänomenologie die Beschreibung dieser Lebenswelt selbst zum Ausgangspunkt gewählt hat, um von diesem aus das Streben nach Exaktheit, Objektivität und dem Ansich der Naturwissenschaft verständlich machen zu können und ihren Anspruch aufzuklären.

Mit der Verteidigung der Geisteswissenschaften, nämlich dass diese auf konkrete Anschauung gerichtet sein sollten, verteidigt die Phänomenologie also sich selbst und ihre eigene Stellungnahme in Beziehung auf das Urrecht der Gegebenheiten der Anschauung, das original Anschauliche als Quelle allen Rechtes.

Dies führt sogar dazu, den Geisteswissenschaften den Naturwissenschaften gegenüber eine Priorität zuzukennen, und zwar darum, weil sie und nicht die Naturwissenschaften in die philosophischen Tiefen führen würden, welche die Tiefen des Letztseienden seien (*Husserliana*, IV, S. 366). Sie sind ja nach Husserl die Wissenschaften der Subjektivität. Während das Subjektive in den Naturwissenschaften als unwichtig betrachtet und so viel wie möglich ausgeschaltet wird, bekommt das Subjektive bei Husserl gerade den Stempel des Wahrhaftigen im besseren Sinne. Die Geisteswissenschaften haben schon in ihrer anfänglichen auf Anschauung gerichteten Form Wahrheit und müssen auch in ihrer weiteren Entwicklung diese Form der Anschauung behalten. Husserl geht sogar so weit, zu behaupten, dass die Naturwissenschaft nur in das Reich philosophischer Wahrheit aufgenommen werden kann, wenn man sie so interpretiert, dass sie ein 'Entquellendes' sei und ihr die absolute Bedeutung entzieht, die sie zu haben beansprucht, aber nach Husserl nicht besitzt. Ihr Anspruch, das An-sich der Natur zu erkennen, soll in einer transzendentalen Interpretation zurückgenommen und am Ende auch als subjektiv im transzendentalen Sinne verstanden werden. Auch die Naturwissenschaft ist schliesslich ein gemeinsames Unternehmen von

Naturforschenden und kann als solche in ihrem Streben nach Exaktheit und Objektivität nur von der historischen Einstellung erklärt und verstanden werden ('Wir die Naturforschenden').

Dilthey suchte eine deskriptive Psychologie neben der naturwissenschaftlichen als Basis für die Geisteswissenschaften.

Husserl übernimmt diesen Gedanken und biegt seine phänomenologische Grundlegung der Psychologie um zu einer Geistespsychologie, einer Psychologie der personalistischen Einstellung, das heisst einer universalen Psychologie der Subjektivität, die in eine transzendentale Phänomenologie einmündet. Hier wird auch die Objektivität der naturwissenschaftlichen Einstellung, die auf die Untersuchung des An-sich von Natur und Welt gerichtet ist, zurückgenommen und uminterpretiert als ihren Ursprung findend in der Subjektivität der Untersucher.

Damit verbindet sich die Zueignung des Gesichtspunktes von Dilthey, dass die Welt des Menschen eine 'gesellschaftlich-geschichtliche' Welt ist. Dadurch wird die Idee der Historizität ein integrierender Teil der phänomenologischen Analyse. Aber eine offene Frage bleibt, welche Wissenschaften nun genau genommen Geisteswissenschaften seien und welche Methode in ihnen angewendet werde oder die am meisten adäquate sei.

Die personalistische Einstellung ist der leitende Gesichtspunkt der universalen Psychologie, sie dient auch als Grundlage aller Geisteswissenschaften.

Sie verbindet sich hier aber sehr eng mit der historischen Einstellung. Geisteswissenschaften sind nach Husserl deskriptiv-anschauliche Wissenschaften und als solche auch historische Wissenschaften. Die personalistische Einstellung fällt hier deshalb zusammen mit der historischen, wie diese in den philologisch-historischen Wissenschaften von altersher sich entwickelt hat. Dies bedeutet im Vergleich mit Dilthey an sich nichts Neues. Auch dieser Denker hat die Geisteswissenschaften mit historischen Kulturwissenschaften identifiziert. Für sie alle ist die Durchführung der historischen Methode charakteristisch. Nur ist Husserl positiver in der Verwendung des teleologischen Gesichtspunktes besonders in seiner geschichtsphilosophischen Auffassung der Stellung und Bedeutung der europäischen Kultur.

Was aber eine schwache Stelle bei Dilthey war, ist von Husserl nicht aufgehoben, sondern zu einem Problem und einer Unsicherheit geworden. Zwar anerkannte Dilthey die Möglichkeit systematischer Geisteswissenschaften neben den historischen, aber diese hatten nicht seine besondere Aufmerksamkeit. Husserl stellt zwar einerseits die Möglichkeit einer objektiven Geisteswissenschaft der historischen gegenüber und andererseits einer apriori'schen phänomenologischen Wesenswissenschaft, aber das geschieht mehr in der Form einer überraschenden und intrigierenden Fragestellung als einer deutlichen Antwort. Die objektive Geisteswissenschaft beruht auf dem Gedanken, dass die ganze konkrete faktische Welt der Erfahrung im Prinzip ganz und gar idealisierbar und deshalb mathematisch konstruierbar sei. Aber dieses hat zur Konsequenz die Forderung, dass sie über eine weltumspannende Mathematik verfügt. Und andererseits fragt er sich, ob es selbstverständlich sei, dass man eine Erforschung der tatsächlichen Geister in ihrer Historizität zu fundieren sucht auf einer Wesenserkenntnis des Geistes und der Gemeinschaft, in der sie leben, und von daher 'exakte Begriffe' schöpft und exakte, unbedingt gültige Wahrheiten als ideale Pole für alle geistige Tatsächlichkeit (*Husserliana*, VI, S. 306). In einer Fussnote stellt er sich dann die Gegenfrage: "Aber wie sieht hier Wesenserkenntnis der historischen Kenntnis aus?"

Dies alles berührt das Problem der Allgemeinheit. Alle Typologie trägt hier ja den Charakter historischer, nicht strenger, eventuell mathematischer Allgemeinheit. Man kann sich deshalb nicht dem Eindruck entziehen, dass sowohl für Dilthey als für Husserl die Wissenschaften vom Menschen als Menschen mit den historischen Geistes- oder Kulturwissenschaften identisch sind. Diese haben ihr Fundament in einer allgemeinen Wissenschaft des Geistes. Husserl anerkennt zwar daneben eine Psychophysik und sogar eine objektive Geisteswissenschaft, welche mit physischen und mathematischen Methoden rein induktiv arbeiten, aber die Geisteswissenschaften haben dennoch ihr eigentliches Fundament in einer Wissenschaft der Subjektivität, da sie selbst empirische Wissenschaften der historischen Subjektivität sind.

In dieser Beziehung erheben sich viele Probleme, wovon wir hier nur einige, die miteinander zusammenhängen, nennen.

(1) Erstens kann man fragen, ob die Wissenschaften vom Menschen tatsächlich nur und ausschliesslich historische Wissenschaften sind. Stimmt das mit der wachsenden Tendenz in diesen Wissenschaften zu Theorienbildung und im Zusammenhang damit der Objektivierung und Formalisierung überein? Ist dieses Streben ohne weiteres identisch mit Naturalisierung oder bedeutet es nicht vielmehr das Erreichen einer

höheren Stufe der Abstraktion in die Richtung einer strukturalen Analyse, z.B. in der Sprachwissenschaft, der Ökonomie und der Soziologie. Die Gleichstellung der Wissenschaften vom Menschen mit historischen (Geistes)Wissenschaften trägt deutlich den Stempel des 19. und Anfanges des 20. Jahrhunderts. So wurde die Sprachwissenschaft prinzipiell als eine historische Wissenschaft aufgefasst: "Sprachwissenschaft ist Sprachgeschichte" (H. Paul). Mit dem Emporkommen einer allgemeinen Sprachtheorie und einer theoretischen Sprachwissenschaft ist dies ganz anders geworden. Dasselbe trifft zu für die soziale, politische und ökonomische Wissenschaft. Auch für die Kunstwissenschaft, welche früher hauptsächlich als Kunstgeschichte aufgefasst und anerkannt wurde, gilt das nicht mehr so unbedingt. Nur die Rechtswissenschaft und die theologische Wissenschaft wurden im 19. Jahrhundert als dogmatische Wissenschaften davon ausgenommen. Diese wurden aber nur als Wissenschaft anerkannt insoweit sie Rechtsgeschichte und Religionsgeschichte waren.

Die Frage erhebt sich deshalb, was die Aufgabe der Wissenschaftstheorie eigentlich sei. Verhält sich diese nur rein beschreibend zu den angewandten Methoden und Zwecken oder auch normierend oder bietet sie sogar leitende Ideen?

Bei Husserl ist dies nicht zweifelhaft. Man kann schwerlich sagen, dass er eine Beschreibung der tatsächlich ausgeübten Methode gibt. Vielmehr bietet er eine Philosophie des Geistes und der Person im Zusammenhang mit dem Natur- und Weltbegriff. Diese ist in Übereinstimmung mit der Tradition defensiv und metaphysisch. Der Dualismus, den er verteidigt, ist schliesslich der Dualismus von Natur und Geist. Was die Wissenschaften betrifft, so führt dieser zu einem Gegensatz zwischen Wissenschaft der Objektivität und Exaktheit auf der Grundlage eines engeren und erweiterten Naturbegriffes und einer Wissenschaft der Subjektivität auf der Grundlage eines Geistes- und Personbegriffes.

(2) Zweitens verdient die Idee des objektiven Geistes, bei Dilthey ein Versuch, um gerade unter Einfluss von Husserls Kritik dem Psychologismus zu entgehen, mehr Aufmerksamkeit in den historischen Geistes- oder Kulturwissenschaften als bei Husserl selbst der Fall ist. Der Begriff des objektiven Geistes ist Husserl zwar nicht fremd, auch räumt er der Leistung, den Leistungs- und Kulturgebilden eine nicht unwichtige Stelle in seiner Theorie ein, aber bei ihm dominiert der Personalismus in den Geisteswissenschaften so sehr, dass die Idee vom objektiven Geist (Werke,

Kunstwerke, Institutionen, politische Systeme, Sprachgebilde) nur wenig hervortritt. Dies ist verständlich, da Husserl nun einmal in den Geisteswissenschaften Wissenschaften der Subjektivität sieht. Das Streben nach Objektivität und sachlicher Verantwortung in Sprachwissenschaft, Philologie, literarischer und historischer Kritik richtet sich besonders auf diese Seite (Sprachformen, das Werk, das Kunstwerk). Auch in der Geschichte der Philosophie selbst ist es empfehlenswert, in dieser Geschichte nicht nur eine Galerie persönlicher Denkweisen und persönlicher Weltanschauungen zu sehen. Das ist von grosser Bedeutung für die Hermeneutik dieser Wissenschaften.[1]

(3) Drittens ist die Theorie Husserls viel mehr als eine Beschreibung wissenschaftlicher Methode, dadurch dass sie die Funktion von Ideen und Normen anerkennt, welche die positive Wissenschaft überschreiten. Einerseits transzendiert sie diese nicht, sondern zeigt eine Richtung an, in der sie sich schon bewegt, wie z.B. die Idee der unendlichen und wahren Welt als Korrelat der Idee der Weltwissenschaft. Andererseits überschreitet sie die positive Wissenschaft ganz bestimmt, indem sie ethische Ideen darstellt, wie die Idee eines wahren und echten personalen Einzellebens, einer echten Gemeinschaft und echter Menschheit.

Am Ende nimmt sie die Idee einer Allwissenschaft als Naturwissenschaft zurück und spricht einer universalen, personalen Wissenschaft, und damit den Geisteswissenschaften, eine Priorität zu. Letzteres beruht auf der Überlegung, dass alle theoretische Fragestellung schliesslich vom Menschen als lebendiger Person ausgeht und dass alle Wissenschaft eine Unternehmung der Menschheit ist.

Schliesslich wird dies alles auf den absoluten Boden der transzendentalen Subjektivität gestellt, welche das Thema der transzendentalen Einstellung und der Phänomenologie als der universalen, absoluten Geisteswissenschaft ist, in der man nicht von der Welt als vorgegeben ausgeht, sondern diese sowohl in ihrer Erscheinungsweise als in ihrem An-sich, mit Wahrheiten an sich als darauf bezogen, als in dem transzendentalen reinen Bewusstsein konstituiert, enthüllt.

University of Utrecht

ANMERKUNG

[1] Siehe K. Kuypers, 'Die Geschichtlichkeit der Sprache und die Aufgabe der Philosophie' in *Das Problem der Sprache* (herausgegeben von H. G. Gadamer), Wilh. Fink Verlag, München, 1967, S. 387–407.

DALLAS LASKEY

EMBODIED CONSCIOUSNESS AND THE HUMAN SPIRIT

> ... from the ashes of the great weariness, the phoenix of a new inner life of the spirit will rise as the underpinning of a great and distant human future, for the spirit alone is immortal.[1]
>
> EDMUND HUSSERL, *Philosophy and the Crisis of European Man*

I. INTRODUCTION

In the decades since World War II there has been a blending or intermingling of problems of existence and traditional phenomenology to form a new existential phenomenology. In America, it has been this version of phenomenology which has been dominant up to now. The doctrine of the *Lebenswelt*, a conception which emerged in the later works of Husserl, has contributed in large measure to this direction. From this common area of concern philosophers have proceeded in radically different directions and have interpreted phenomenology in strikingly different ways. The older phenomenology, often characterized as the phenomenology of the essences, tried to delineate the invariant and universal structures of human consciousness especially at higher levels. The reduction was the gateway to this phenomenology, for in order to get at data without distortion, it was necessary to "put into brackets" or to disengage the ego from its natural concerns. This was in no way to cut it off from the world; but, unfortunately, critics have taken the reduction to involve an ontological separation, despite Husserl's lifelong denials. Hence traditional phenomenology was criticized as idealistic and as a mere study of consciousness. Some of those who disagreed with this view of phenomenology "tried to bring it down into the world of concrete life and to incarnate it in individual and social human existence."[2] This phenomenology does not deny consciousness, but employs a different vocabulary, the basic unit being the body-subject, whose subjectivity is to be systematically explored. Through the doctrine of the body-subject Merleau-Ponty[3] was

Tymieniecka (ed.), Analecta Husserliana, Vol. I, 197–207. *All Rights Reserved*
Copyright © 1970 *by D. Reidel Publishing Company, Dordrecht-Holland*

able to show the contribution of our bodies as sense-bestowing intentionality. In this perspective is not so much the distinguishing mark of consciousness, as a basic characteristic of the body, acting and being acted on in the world. Merleau-Ponty systematically reinterprets nearly every technical term of the Husserlian phenomenology. Even the nature of rationality itself becomes altered.

While existential phenomenology has attracted many followers in America, transcendental phenomenology, with a few exceptions, stands strangely alone. Yet it may have been dismissed too readily, perhaps because it was never really understood. This is not to deny the real difficulties it faces in the problems of the other self and the constitution of the body. Nevertheless, I am not yet convinced that such difficulties in transcendental phenomenology are any less formidable than the task of finding support for and clarification of the fundamental ontological theses required to get existential phenomenology underway. On behalf of transcendental phenomenology, I wish merely to affirm that it seems especially suitable for the treatment of certain areas of human experience; notably, those higher reaches of the human spirit which have traditionally claimed universality and autonomy.

In the opposition between the phenomenology of Husserl and of Merleau-Ponty the problem of the constitution of the body plays a central role. In particular, existential phenomenology has been profoundly influenced by the doctrine of the body-subject and the reduction of existence to being-in-the-world. It is the interpretation provided by Merleau-Ponty that has attracted so much attention in America. In this paper, I hope to show that the ontological theses involved in the doctrine of the body-subject (1) go beyond any support advanced for them, (2) involve elements which threaten the very nature of higher consciousness and its claim for autonomy and (3) prevent the 'rebirth of Spirit' (as Husserl speaks of it in the *Crisis* article) which was necessary to overcome the crisis facing western civilization. To accomplish this objective, I propose first to provide a perspective for the discussion by a consideration of two central themes which appear in Husserl's *Crisis* articles; namely, his view of a new conception of rationality, and the idea of a universal science of the human spirit. When this perspective is delineated, the question of the constitution of the body is then studied. Husserl's solution of the problem permits the establishment of autonomous levels of the human spirit. Merleau-Ponty's solution not only fails to do this but involves serious ambiguities in understanding and interpretation. Some of the existential elements contributing to the confusion are examined. Finally, reasons are advanced for a reconsideration of Husserl's program of transcendental phenomenology insofar as the latter appears to account for a quality of human existence that existential phenomenology cannot or will not.

II. THE REBIRTH OF THE HUMAN SPIRIT

Husserl's *Crisis*[4] will remain one of the more challenging works of the 20th century, not only because of its bearing on the development of his own phenomenology, but because it provides themes for alternative and opposing conceptions of phenomenology. In any discussion of the differences between these two trends of phenomenologies it is useful to look at the *Crisis* in order to clarify the original prospects and tasks indicated there. It also provides a number of common themes which serve as a ground for further comparison.

What was the crisis? Are there aspects which have been overlooked or insufficiently studied? An affirmative answer is clearly indicated. For Husserl the crisis was the "seeming collapse of rationalism". This affected the very pattern of our own existence with loss of meaning and uncertainty of one's own mission or destiny.[5] When norms, standards and ideals lost their meaning, a new barbarism threatened man. Now existential phenomenologies have accepted the 'collapse of rationalism' seemingly as a fact; a review of the works of Heidegger, Sartre and Merleau-Ponty would lead one to the conclusion that rationalism is not only dead but buried for all time. However, it is one thing to affirm this and quite another to demonstrate it. What kind of rationality was it that was on the verge of collapse? For Husserl it was not the rationality of the classical Greeks but a distortion of it, a one-sided version which appeared after the 18th century and which resulted in a split between man and nature. When the world of nature was set off from man and described in impersonal and mathematical terms, a new kind of objectivity and 'naturalism' arose which had harmful repercussions for man's subjectivity. It was this one-sided or partial rationalism that was taken universally to apply to both nature and spirit. The entire content of conscious life was in danger of being objectivized in the same way nature had been objectivized. The whole of cultural life was presumed to be capable of rational explanation in objective and scientific categories. Man's very existence and behaviour were considered as *effects* following from antecedent factors in such a way as to permit formulation in scientific terms, i.e. subsumed under universal law. His loves, hopes, aspirations, and even his reasoning itself seemed capable of being brought under the rule of scientific explanation. But with this naturalizing of the human spirit[6], purposes, values and norms lost their claim for universality and necessity. Husserl believed that when man *loses the ideal norms for infinite tasks, he loses the source of spiritual renewal.* It is the conception of infinite tasks and infinite ideas which have been the source of enormous creativity in western culture, and the absence of which leads to questionable existence and threadbare purpose and meaning. The path through relativism, skepticism, cynicism and nihilism becomes all too clear.

The one-sided version of rationality had split off the realm of nature from the world as it is in fact lived and experienced. This meant that the scientist dealt with nature as a *self-contained world* and was able to separate out all the human contributions; hence, he could consider nature purely as nature.[7] The humanist was *not* able to achieve the same kind of separation and autonomy for the realm of spirit and culture, because of the belief that there is a causal tie between the corporeal and spiritual existence such that any explanation of spirit was ultimately forced back to bodily manifestations – and an explanation in terms of physics and chemistry would then be appropriate. A science of the human spirit became impossible. *Now it is just this mode of thought that Husserl condemns!*

Here then was the crisis. What could be done about it? Only a revitalized conception of rationality would restore dignity and value to human existence. The means he then proposed was transcendental phenomenology. Of all the philosophies then extant, only transcendental phenomenology could do justice to the particularly human themes; in other words, it was the only way to avoid the kind of objectivity which was destroying man's humanity.

The main issue is the failure to develop a proper science of the human spirit because of the excesses of naturalism and objectivism. Husserl indicates the root of the mistake as the inability of the natural sciences, blinded by their own naturalism, to see the possibility of a universal and pure science of the spirit.[8] What is required to solve the problem is a new conception of reason to replace the old; it is a conception of spirit [9] understanding itself in a really universal and radical manner wherein an "entirely new scientific thinking is established in which every conceivable question, whether of being, norm, or of so-called 'existence' finds its place."[10] Husserl looked for a total transformation of knowledge in this new realm of spirit which embraced "all being in an absolute historicity" into which nature fits as a product of spirit. Thus transcendental phenomenology is called upon to make spirit as spirit the field of systematic investigation.

It was such a conception that Husserl had in mind when he talked of "the rebirth of spirit". Man is the rational living being who has reached a new level of existence beyond the animal and the savage. On this level of human existence with its ideal norms and infinite[11] tasks, the relativist and historicist views are transcended. Husserl is thinking of a rationality, not as bounded and limited by particular social and cultural contexts, but as *absolutely universal* in its cognitive claims and demands.

It is clear that such a conception of rationality is not to be found on the level of the *Lebenswelt*, but on the contrary, on the higher level of human consciousness. Science and philosophy, to be taken seriously, must make a universal demand. Husserl seemed to believe that until universal norms[12] and infinite tasks were articulated capable of sustaining man and giving him a sense of mission and purpose, the crisis of civilization would threaten man.

It is in the context of this crisis of civilization and the rebirth of the human spirit that I wish to raise the question of the constitution of the body. Does the proposal by the existential phenomenologists contribute to the solution of these questions, or does it introduce paradoxes and complexities?

III. CRITIQUE

In the preceding part the main concern was to present a perspective for a fresh look at the two phenomenologies. Now the task is critical. The one guiding principle for this part of the investigation is borrowed from Husserl's principles of all principles: "That whatever presents itself in 'intuition' in primordial form (as it were in its bodily reality), is simply to be accepted as it gives itself out to be ..."[13] In other words, all statements require evidence and this evidence must be given immediately in experience. Statements failing to meet this demand will not be considered. Now if this requirement is accepted, then a number of fundamental statements in Merleau-Ponty's existential phenomenology are suspect. I would like to consider the following two statements in turn: (1) "The world is there before any possible analysis of mine ..."[14] and (2) "The phenomenal world is not the bringing to explicit expression of a pre-existing being, but the laying down of being".[15]

How is one to interpret the statement "The world is there before any possible analysis of mine ..."? What evidence can be produced for this? These are questions that do not permit an easy answer. Merleau-Ponty has stressed again and again in *Phenomenology of Perception* that the world is not to be taken as Absolute Spirit in the idealist sense nor as an independent reality in the realist sense. He also affirms that the world is the natural setting for all that takes place; it is in such a world that he wishes to insert human existence, so as to avoid the possibility of a Cartesian split between consciousness and nature. Throughout his book he affirms the world is not an object, is not a consciousness, is not an inner state. Yet in some manner or other it is 'already there'. How can one interpret this? If it is affirmed at the level of lived experience, then it would have no ontological or existential significance. On the other hand, to dissociate his philosophy from the so-called 'idealism' of Husserl, he must intend his statement as something more than a description of a conscious state. Some sort of being is obviously intended for the world as a natural setting, but it is not at all clear. The implication would seem to be that this vast context is somehow real and at the same time existent. One is forced to assume that "the world is already there" is an ontological assertion made in order to get his philosophical enterprise underway.

The problem just indicated seems to be a problem for any form of existential phenomenology which would break away from Husserlian phenomenology. Part of the trouble comes from the way the *Lebenswelt* has been interpreted. Some philosophers[16] have been led to take it as some kind of fundamental metaphysical or ontological ground for the development of a phenomenology of a more concrete nature. Once the *Lebenswelt* is taken in a metaphysical sense, then phenomenology can simply read off its structures as structures of the world. Similarly, it becomes increasingly attractive to explore the *Lebenswelt* to the preconscious, preobjective stratum as a fundamental source or ground. While traditional phenomenological language is used in this enterprise, it is not clear what precisely is meant. This is not unexpected, of course, when one departs from the controls of traditional phenomenology. Without a clearly executed reduction, it is simply difficult to know the level of discourse or experience.

Husserl found it quite easy to speak of the existence of the world when in the natural standpoint. However, by means of the *epoché* he was able to transfer to the phenomenological standpoint, Merleau-Ponty does not really have a reduction in the sense of Husserl; while it is sometimes held that the account of the pathological examples serves as a sort of *epoché* to redirect attention to the lived experience, this does not appear very convincing. Nor is the merely making our concern the object of philosophical inquiry sufficient to grasp the lived experience in its purity. It should be pointed out that for Husserl the *Lebenswelt* no more exists than does the scientific world; the scientific world is the intentional correlate of scientific intendings; the *Lebenswelt* is the correlate of lived experience.

Part of the difficulty and obscurity which abounds in Merleau-Ponty's *Phenomenology of Perception* comes from these ontological questions. If the book is anything more than a basic description of our lived experience, then how is it to be taken? Is he advocating a correct ontology as opposed to the false ontology of objectivism? If this is denied, then how can the existential claims of the book be justified? If the question is affirmed, then the book is claiming more than can be justified through phenomenological descriptions. Some kind of ontological commitments seem to have been made before the enterprise gets underway; but they pose seemingly insoluble problems.

A further note may be added concerning the matter of prior ontological commitments. I wish to advance the hypothesis that Merleau-Ponty is committed in advance to the belief in the world as not completely rational. In other words, the world contains surd elements, elements which by their very nature resist all attempt at rational explanation. Can this hypothesis be supported? Two points immediately come to mind. The first concerns the direction of his investigations. Instead of systematically exploring the evidence for rationality in the areas you would most likely expect to find it (the natural sciences, logic, mathematics, law, etc.) he goes below the level of lived experience to that of the preobjective, preconscious where one would least expect to find it. It, therefore, comes as no surprise that he encounters non-rational elements. The second point concerns the locale of rationality; this is not found in the reaches of higher consciousness, but in the world.[17] This is not the world in the realist sense, however. Merleau-Ponty claims that "rationality is precisely measured by the experiences in which it is disclosed. To say that there exists a rationality is to say that perspectives blend, perceptions confirm each other, a meaning emerges."[18] It seems to be identified with the sphere of action and conduct:

> But the meditating Ego, the 'impartial spectator' (*uninteressierter Zuschauer*) do not rediscover an already given rationality, they 'establish themselves', and establish it, by an act of initiative which has no guarantee in being, its *justification resting entirely on the effective power which it confers on us of taking our own history upon our selves.*[19] (my italics)

Here the cat is clearly out of the bag. Rationality is established by a daring act of initiative whose justification rests ultimately on the power conferred on the person making the choice. Here the old structures of rationality in the intellectualist sense are replaced by the new volitional structures of our affective life. With this substitution is a change from *criterion of logic* to a *criterion of power*. Though the conception of rationality seems adjusted to the contemporary mood of revolutionary change, and one which would easily justify the revolutionary violence that has now appeared on the scene, one must admit that it opens the door to all kinds of future tyranny and arbitrariness. In my opinion, this is no alternative at all to meet the crisis of civilization.

The ontological issue is equally a problem for Merleau-Ponty's most fundamental concept, that of the body-subject. It is by means of this that body and consciousness are unified and the *Cartesian* dualism overcome. This is not a psyche joined to a subject[20] but a fundamental unity. Now the body-subject has the same sort of status trouble that the world had because it is inextricably involved with it. It is by means of the body-subject that we are inserted into a world; hence, if there is uncertainty about the concept 'world' there is equal uncertainty over the body-subject.

At the level of lived experience nothing is more certain than the unity of man's existence. But when we are talking about lived experience we are talking about a *phenomenon* and not a *being*. Yet Merleau-Ponty attributes characteristics to the body-subject that seem to run counter to its status; for example, the body-subject becomes a source of meaning for it is the center of action. It is an intentional action of such a sort that objects in the world become a kind of intentional correlate of bodily activity. The body-subject now serves as a sort of existential substitute for consciousness; it now has to do two things, (1) to assist in the constitution of objects, (2) to know such objects – but these two functions do not go together, for the body-subject is an instrumentality for the I, though it is not identical with it.

What is most significant about the theory of the body-subject is that so many non-conscious factors enter into its constitution. The meanings we find in our world are not created by us, but are organized by the body which is in a kind of dialogue with the world by means of a new kind of two-way corporeal intentionality. The evidence for this assertion is not provided directly, for he is dealing with what lies beyond consciousness. Intentionality has become an ontological relationship and is, when pushed to its depths, preconscious. What used to be considered a cognitive relationship is now a relationship of being. One wonders at this point what happens to the phenomenology; all of the fundamental terms are quietly and unobtrusively altered from the phenomenal level to the level of being.

All experience whatsoever is construed on the model of perceptual experience, for the theory of the body-subject is a theory of perception. All our experiences at the cultural and intellectual level are now accounted for in terms of a new kind of constitution wherein non-conscious elements play an important role. It is evident that Merleau-Ponty is more concerned with what is prior to consciousness than with its higher reaches:

> ... consciousness must be faced with its own unreflective life in things and awakened to its own history which it was forgetting; such is the *true part* that philosophical reflection has to play.[21] (my italics)

Has he not given up the possibility that the higher planes of conscious life are autonomous and have their own kind of order? Rather than examining the order and contents of higher consciousness as they present themselves immediately in our experience, he is led to account for them in genetic fashion from their own origin. Such origins being beyond the range of the conscious are obscure and ambiguous. Now this is a curious position to take, especially when the contrary evidence is so close at hand. Looking at science as a cultural achievement one can provide a phenomenological account of its origins. In the standpoint of science itself, however, the order, precision and rationality of its concepts and methods are available and open to the inspection of all. Similarly, within the standpoint of logic one can demonstrate the autonomy of logical relations. It is one thing to show how fundamental logical ideas grew out of social and historical contexts, but this has nothing to do with the formal relations between such logical structures themselves. Merleau-Ponty has presented a curious theory which is reductivist in a strange non-causal way. From the murky depths of the preconscious the surd elements are retained so as to destroy the very concept of autonomy and universality. Norms, values and ideas are all charged with a limitation and a situatedness which is built into the theory. It is no accident that Merleau-Ponty did not develop an ethical theory or theory of value; it is difficult to see how this conception of experience could serve as a base for the development of the new humanity. By contrast with Husserl it offers almost a deliberate choice of the irrational.

In the wider context of Husserl's *Crisis* and the rebirth of the human spirit, the concept of the body-subject is unable to provide for the universality and autonomy that is revealed at the higher levels of the spirit. In those cultural achievements where rigor, elegance and precision are the distinguishing marks, the theory of the body-subject can only offer relativism and historicism. In no way can this doctrine meet the challenge indicated by Husserl. Philosophy, when taken seriously, must make a claim beyond the individual and the particular. If it remains at a low level of generality, then it becomes increasingly difficult to judge its claims. It is the ultimate cognitive significance which is threatened by the move from the traditional conceptions of phenomenology to its more ontological versions.

REFERENCES

[1] Edmund Husserl, *Ideas* (transl. by W. B. R. Gibson), George Allen & Unwin Ltd., London, 1958, pp. 101–111.

[2] Herbert Spiegelberg, *The Phenomenological Movement*, Vol. 2., Martinus Nijhoff, The Hague, 1960, p. 559.

[3] M. Merleau-Ponty, *Phenomenology of Perception*, Part I (transl. by Colin Smith), Routledge & Kegan Paul, London, 1967.

[4] Edmund Husserl, 'Philosophy and the Crisis of European Man', in Quentin Lauer, *Phenomenology and the Crisis of Philosophy*, Harper Torchbooks, New York, 1965.

[5] Husserl, *Crisis.*

[6] *Ibid.*, p. 190.

[7] *Ibid.*, p. xxx.

[8] *Ibid.*, p. 155.

[9] Scheler who develops this idea further, claims that the act of spirit (ideation) is

completely free from all technical intelligence and from inferential thought. (*Man's Place in Nature*, Noonday, New York, 1962, pp. 49–50.)

[10] *Crisis*, p. 190.

[11] *Ibid.*, p. xxx.

[12] Husserl speaks of norms and ideas, not as things which are in space, but as factors relevant to the human being as a person. It is the presence of great ideas spreading through a community and a mode of society which activated the spiritual and intellectual attainment.

[13] Edmund Husserl, *Ideas* (transl. by W. R. Boyce Gibson), George Allen & Unwin Ltd., London, 1958, p. 92. Also *Cartesian Meditations*, p. 13.

[14] M. Merleau-Ponty, *Phenomenology of Perception* (transl. by Colin Smith), Routledge & Kegan Paul, London, 1967, p. xi.

[15] *Ibid.*, p. xx.

[16] Gerhart Funke, *Phänomenologie – Metaphysik oder Methode*, H. Bouvier & Co. Verlag, Bonn, 1966, p. 139.

[17] *Phenomenology of Perception*, p. 430.

[18] *Ibid.*, p. xix.

[19] *Ibid.*, p. xx.

[20] *Ibid.*, p. 88.

[21] *Ibid.*, p. 31.

Private v. Public ϕ - 13
Eidetic Reduc - 13ff
H as metaphysician - 16
T-ϕ - 17f
Turtleism - 18
Sci - 18